Junshan Zhang

Physical-layer Aware Control and Optimization in Wireless Networks

Dong Zheng
Junshan Zhang

Physical-layer Aware Control and Optimization in Wireless Networks

VDM Verlag Dr. Müller

Impressum/Imprint (nur für Deutschland/ only for Germany)
Bibliografische Information der Deutschen Nationalbibliothek: Die Deutsche Nationalbibliothek
verzeichnet diese Publikation in der Deutschen Nationalbibliografie; detaillierte bibliografische
Daten sind im Internet über http://dnb.d-nb.de abrufbar.

Coverbild: www.purestockx.com

Verlag: VDM Verlag Dr. Müller Aktiengesellschaft & Co. KG
Dudweiler Landstr. 99, 66123 Saarbrücken, Deutschland
Telefon +49 681 9100-698, Telefax +49 681 9100-988, Email: info@vdm-verlag.de
Zugl.: Tempe, Arizona State University, Diss., 2007

Herstellung in Deutschland:
Schaltungsdienst Lange o.H.G., Berlin
Books on Demand GmbH, Norderstedt
Reha GmbH, Saarbrücken
Amazon Distribution GmbH, Leipzig
ISBN: 978-3-639-13434-6

Imprint (only for USA, GB)
Bibliographic information published by the Deutsche Nationalbibliothek: The Deutsche
Nationalbibliothek lists this publication in the Deutsche Nationalbibliografie; detailed
bibliographic data are available in the Internet at http://dnb.d-nb.de.

Cover image: www.purestockx.com

Publisher:
VDM Verlag Dr. Müller Aktiengesellschaft & Co. KG
Dudweiler Landstr. 99, 66123 Saarbrücken, Germany
Phone +49 681 9100-698, Fax +49 681 9100-988, Email: info@vdm-publishing.com
Tempe, Arizona State University, Diss., 2007

Printed in the U.S.A.
Printed in the U.K. by (see last page)
ISBN: 978-3-639-13434-6

To our families.

ACKNOWLEDGMENTS

I am greatly indebted to my advisor, Professor Junshan Zhang, for his continuous support and close guidance. I really appreciate his patience and encouragement while I was searching for my Ph.D. thesis topic. His enthusiastic attitude towards research quality always drives me to work the best of mine. His willingness and accessibility to discuss technical and non-technical issues with me made my student life much easier and more enjoyable. I am very fortunate to have him as my advisor, who put equality between an advisor and a friend for me.

I would like to thank Professor Joseph Hui for teaching me broadband networks and queuing theory. His master skills in these topics are truly remarkable. I learned most of my coding and information theory knowledge from Professor Tolga Duman. His classes are very enjoyable since he can always explain any complicated concepts in a much simpler way. Thanks also go to Professor Martin Reisslein and Professor Antonia Papandreou-Suppappola for serving in my doctoral committee and for their valuable suggestions on my thesis.

I would like to express my gratitude to Professor P.R. Kumar for accommodating me during my winter stay in UIUC. Thanks for giving me such a wonderful chance to know many outstanding researchers there. I benefited a lot from the game theory course taught by Professor Tamer Başar and the stochastic control theory course taught by Professor Kumar.

Many thanks to my current and former colleagues: Ming Hu, Bo Wang, Ning He, Ping

Gao, Qian Ma, Qinghai Gao, Zhifeng Hu, Chandrashekhar Thejaswi, Shanshan Wang, Min Cao and Vivek Raghunathan for the pleasant and inspiring discussions. I am also fortunate to know some good friends here who always support me in one way or another; they are: Qiang Zhan and Jingna Xia, Weiyan Ge and Yiran Zheng, Kai Bai and Xun Sun. Their friendship is a valuable experience for me.

Finally, I want to thank my parents Riyong Zheng and Rong Cheng, my parents-in-law Zunqi Liu and Baiyun Yi, and my wife Xuhong Liu for their endless love and support. This thesis is dedicated to them!

iv

II   Utility-based Random Access and Flow Control   115

LIST OF TABLES

CHAPTER 1

INTRODUCTION

The age of ubiquitous communications and computing is upcoming, and it is changing dramatically the way we manage, search, and access information. With the explosive growth in demand for wireless services, the next generation wireless networks are expected to provide solutions to cost-effective, ubiquitous, always-on broadband access. Indeed, the wireless technology evolution is taking place in many fronts: multiple antenna techniques, modulation/coding schemes, medium access control (MAC) design, routing protocols and security mechanisms. There has been a general consensus that developing network-level solutions that take advantage of the interplay between the physical-layer and the upper protocol layers would yield significant performance gains. Thus motivated, a common thread of this dissertation is to study physical-layer aware control and optimization in stochastic wireless networks.

To exploit the physical-layer (PHY-layer) properties for upper-layer protocol design, it is of critical importance to obtain a fundamental understanding of the following issues: Is it possible to make use of channel fading for the MAC design, and if yes, how? How can we take advantages of rich diversities such as time/multi-user/multi-channel/multi-receiver degrees of freedom in wireless networks? Is it necessary to modify the networking protocols (such as MAC), and if yes, how can we integrate the new features of the PHY-layer into the MAC design? How can we devise MAC protocols towards link-level QoS provisioning in wireless networks in the presence of channel variation and topology-dependent user contentions? Furthermore, how can we develop distributed algorithms of joint rate control and MAC design for end-to-end QoS provisioning? Are these distributed algorithms stable in the presence of stochastic noisy feedback? A principal objective of this dissertation is to develop systematic frameworks for solving these issues.

In the following, we first give an overview of stochastic wireless networks.

1.1. Scope of The Dissertation

The traditional design methodology of wireless networks follows the standard OSI layered structure of wireline networks [87]. The studies in this dissertation explore the optimization and control across the following three layers (see Fig. 1.1):

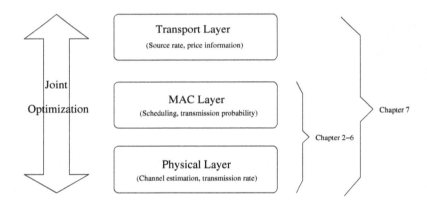

Figure 1.1. A layered view of wireless networks.

Physical (PHY) layer: A main functionality of the PHY-layer is to transmit bit streams reliably over communication links. In particular, the studies in this dissertation involve channel estimation, modulation/coding and rate selection.

Medium access control (MAC) layer: The MAC layer is to allocate channel resources among competing links in the network. The sharing of the channel can be by time-division, frequency-division, code-division, or a combination of these.

One underlying model for our studies is random access [14]. The design parameters are the link scheduling and the transmission persistence probabilities.

Transport layer: The transport layer transfers data packets between end systems, and is responsible for end-to-end error recovery and congestion control. In this study, we consider how to generate pricing information according to traffic congestion levels along the path, and how to perform source rate adaptation based on the generated pricing information under fairness constraints.

Throughout this dissertation, we explore joint optimization and design of these layers by taking into account of the PHY-layer properties into upper layer protocol design. In the following, we give a brief description of the studies in each chapter.

1.2. Summary of Main Contributions

Roughly speaking, the main body of this dissertation can be categorized into two thrusts. In the first thrust (Chapters 2-5), we develop channel-aware distributed scheduling for exploiting rich diversities in MAC design, and we take a novel approach to formulate wireless scheduling for throughput maximization problem, using a stochastic decision making framework. In the second thrust (Chapters 6-7), we take a network utility maximization (NUM) approach to study MAC design, joint flow control and random access in wireless ad hoc networks and examine the stability therein. Although the application of the NUM framework in network control design is not new (for example, the recent advances of TCP forward and reverse engineering, e.g., [63, 62, 64]), its application to wireless ad hoc networks, especially to random access networks, has appeared only recently (e.g., [56, 94, 88]). In addition, our work is based on the recent process on stochastic approximation for constrained optimization and rate of convergence. A brief summary of the main contributions in each chapter in the rest of the thesis is given below.

1.2.1. Channel-aware Distributed Scheduling

In Chapters 2-4, we consider distributed opportunistic scheduling (DOS) in wireless ad-hoc networks, where many links contend for the same channel using random access. In such networks, DOS involves a process of joint channel probing and distributed scheduling. Due to channel fading, the link condition corresponding to a successful channel probing could be either good or poor. In the latter case, further channel probing, although at the cost of additional delay, may lead to better channel conditions, and hence yield higher transmission rates. The desired tradeoff boils down to judiciously choosing the optimal stopping rule for channel probing and the rate threshold.

In Chapter 2, we pursue a rigorous characterization of the optimal strategies from a network-centric perspective where links cooperate to maximize the overall network throughput. Using optimal stopping theory, we show that for a homogenous network, the optimal scheme for distributed opportunistic scheduling turns out to be a *pure threshold policy*, where the rate threshold can be obtained by solving a fixed point equation. We then generalize the above study to the heterogeneous case. Somewhat surprisingly, the optimal rate threshold turns out to be the same across all links regardless of the channel statistics and transmission probabilities. We further devise iterative algorithms for computing the threshold. We also generalize the studies to take into account fairness requirements and time constraint.

In chapter 3, we generalize the above study to the case where the channel condition is estimated using noisy observation. We show that the transmitted rate has to back off on the estimated rate so as to reduce the channel outage probability. Furthermore, the optimal scheduling policy is still threshold-based; the optimal threshold turns out to be a function of the variance of the estimation error, and indeed, it is a functional of the backoff rate. Due to the intractability of the optimal backoff, we propose a suboptimal scheme that backs off on the estimated SNR (and hence the rate) through a backoff ratio σ. And we propose an iterative algorithm to compute the optimal backoff ratio and the optimal threshold. Simulation results are provided to show that the corresponding distributed opportunistic scheduling algorithm still achieves significant throughput gain in the presence of noisy channel estimation, especially in the low SNR region.

Chapter 4 generalizes the study of single channel/receiver systems in Chapter 2 to multi-channel/multi-receiver systems. Specifically, in such networks, we show that channel probing takes place in two phases: 1) in phase I, each transmitter contends for the channel using random access to reserve the channel, and the probing to accomplish a successful channel contention takes a *random duration*; and 2) in phase II, subsequent probings are carried out to estimate the link conditions to different intended receivers of this successful transmitter, according to specific probing mechanisms, and the probing for each receiver takes a *constant duration*. In particular, we shall study four probing mechanisms for utilizing multi-receiver diversity and multiuser diversity, namely, 1) the random selection (RS) mechanism, 2) the exhaustive sequential probing with recall (ESPWR) mechanism, 3) the sequential probing without recall (SPWOR) mechanism, and 4) the sequential probing with recall (SPWR) mechanism. We show that the optimal scheduling policies for all four probing mechanisms exhibit threshold structures, indicating that they are amenable to easy distributed implementation. In particular, it turns out that for RS and ESPWR, the optimal scheduling strategies are single threshold policies, whereas those for both SPWOR and SPWR are multi-stage threshold policies. We show that the optimal thresholds and the maximum network throughput can be obtained off-line by solving fixed point equations. We further develop iterative algorithms to compute the optimal thresholds and the throughput.

Different from the study in the previous chapters where the optimal scheduling policy is derived based on a mathematical framework, in Chapter 5, we study practical MAC protocol design for multichannel wireless networks. Specifically, building on the IEEE 802.11 standard, we propose an opportunistic multi-channel MAC protocol (OMC-MAC), with three key features: 1) by exploiting the channel variations across multiple channels, OMC-MAC achieves selection diversity gain in an opportunistic and distributed manner; 2) the size of the contention window is adjusted adaptively based on the estimate of the number of competing stations, which is obtained via using a Sequential Monte Carlo technique; and 3) OMC-MAC achieves "resource pooling" and thus improves the stability of the network. Analysis results reveal that OMC-MAC in wireless LANs achieves significant throughput gain, even under heavy traffic conditions. Extensive simulation studies show that OMC-MAC can achieve efficient channel utilization for each added channel, compared with the standard 802.11 MAC protocol and other multichannel protocols such as DCA and MMAC [95, 84]. Finally, we show via examples that the sequential Monte Carlo method is effective for the adaptation of the contention window size.

1.2.2. Utility-based Random Access and Flow Control

In Chapter 6, we take a utility maximization approach to study fair MAC design towards QoS provisioning. To this end, we first identify two key challenges of wireless access control, namely the topology dependency and the channel dependency therein. Based on the observation that the topology change and channel variation occur on different time scales, we decompose the utility maximization to two phases: a *"global" optimization phase* addresses the topology dependency, and arbitrates fair channel access across the links by adapting the persistence probability to achieve long-term fairness, and a *"local" optimization phase* deals with the channel dependence, and determines the transmission duration based on local channel conditions while maintaining short-term fairness. Observing that the MAC throughput depends on the realizations of channel contention in random access networks, we use stochastic approximation to investigate in depth the MAC design with the adaptive persistence mechanism in the global phase. Using Lyapunov's Stability Theorems and LaSalle's Invariance Theorem, we establish the stability of the proposed algorithm for the global phase and analyze the fairness under $(\vec{w}, \kappa)$-fair utility functions. Our findings reveal that under the large network assumption, there exists a single equilibrium point for the proposed $(\vec{w}, \kappa)$-fair MAC algorithm provided that $\kappa > 1$. We also present the solution to the local optimization phase under general fairness constraints.

Observing that the implementation of distributed network utility maximization (NUM) algorithms hinges heavily on information feedback through message passing among network elements. In practical systems the feedback is often obtained using error-prone measurement mechanisms and suffers from random errors. In Chapter 7, we investigate the impact of noisy feedback on distributed NUM.

We first study the distributed NUM algorithms based on the Lagrangian dual method, and focus on the primal-dual (P-D) algorithm, which is a single time-scale algorithm in the sense that the primal and dual parameters are updated simultaneously. Assuming strong duality, we study both cases when the stochastic gradients are unbiased or biased, and develop a general theory on the stochastic stability of the P-D algorithms in the presence of noisy feedback. When the gradient estimators are unbiased, we establish, via a combination of tools in Martingale theory and convex analysis, that the iterates generated by distributed P-D algorithms converge with probability one to the optimal point, under standard technical conditions. In contrast, when the gradient estimators are biased, we show that the iterates converge to a contraction region around the optimal point, provided that the biased terms are asymptotically bounded by a scaled version of the true gradients. We also investigate the rate of convergence for the unbiased case, and find that, in general, the limit process of the interpolated process corresponding to the normalized iterate sequence is a stationary reflected linear diffusion process, not necessarily a Gaussian diffusion process. We apply the above general theory to investigate stability of cross-layer rate control for joint congestion control and random access.

Next, we study the impact of noisy feedback on distributed two time-scale NUM algorithms based on primal decomposition. We establish, via the mean ODE method, the convergence of the stochastic two time-scale algorithm under mild conditions, for the cases where the gradient estimators in both time scales are unbiased. Numerical examples are used to illustrate the finding that compared to the single time-scale counterpart, the two time-scale algorithm, although having lower complexity, is less robust to noisy feedback.

Part I

Channel-aware Distributed

Scheduling

CHAPTER 2

DISTRIBUTED OPPORTUNISTIC SCHEDULING: A UNIFIED PHY/MAC

APPROACH

2.1. Introduction

2.1.1. Motivation

Wireless ad hoc networks have emerged as a promising solution that can facilitate communications between wireless devices without a planned fixed infrastructure. Different from its wireline counterpart, the design of wireless ad hoc networks faces a number of unique challenges in wireless communications, including:

1) *Co-channel interference* among active links in a neighborhood: The shared nature of wireless medium may result in transmission failure due to co-channel interference from other transmissions. Collision resolution and interference management are regarded as the functionalities of the medium access control (MAC) layer, and are typically handled by scheduling or random access protocols.

2) *Time varying channel conditions over fading channels*: Fading is the time variation of the wireless channel due to two effects: large-scale path loss and shadowing effects that cause the signal to attenuate with distance; and multipath scattering effects that result in delayed copies of the signal adding up constructively or destructively at the receiver. Fading is often mitigated at the physical layer using coding/modulation and diversity techniques.

The traditional approach for wireless network design intends to separate packet losses caused by fading from those by interference. That is, the PHY layer addresses fading,

[1]©[2009] IEEE. Reprinted, with permission, from Dong Zheng, Weiyan Ge and Junshan Zhang, "Distributed Opportunistic Scheduling For Ad-Hoc Communications: An Optimal Stopping Approach", IEEE Trans. On Information Theory, Vol. 55, Issue 1, Jan. 2009 Page(s):205-222.

while the MAC layer addresses the issue of contention. This hope for separation of point-to-point link reliability and multiple access functionality between the PHY and MAC layers relies on the implicit assumption that the PHY layer works perfectly and hides fading from MAC. However, it is difficult to determine if packet losses are due to MAC-layer variation or channel variation. Indeed, as shown in [4] and our experimental measurements [28], fading can often adversely affect the MAC layer in many realistic scenarios. In general, channel fading and interference occur on the same time scales.

The coupling between the timescales of fading and MAC calls for a unified PHY/MAC design for wireless ad-hoc networks, in order to achieve greater operational efficiencies vis-a-vis throughput and latency. Indeed, joint PHY/MAC diversities, including multiuser diversity, multi-receiver diversity, time diversity and spatial diversity, are available for exploitation in a wide range of wireless scenarios by using channel-aware scheduling. It is therefore of critical importance to develop a rigorous understanding of MAC-layer scheduling that can resolve contention and mitigate interference efficiently while exploiting diversities. There is a consensus that channel-aware distributed scheduling will help to propel significant advances towards providing better QoS in ad hoc networks.

Notably, there has recently been a surge of interest in channel-aware scheduling and channel-aware access control. Channel aware opportunistic scheduling was first developed for the downlink transmissions in multiuser wireless networks (see, e.g., [3], [6], [24], [54], [60], [82], [93]). Opportunistic scheduling originates from a holistic view: roughly speaking, in a multiuser wireless network, at each moment it is likely that there exists a user with good channel conditions; and by picking the instantaneous "on-peak" user for data transmission, opportunistic scheduling can utilize the wireless resource more efficiently. *A key assumption in these studies is that the scheduler has knowledge of the instantaneous channel conditions for all links, and therefore the scheduling is centralized.*

Channel aware random access has been investigated for the uplink transmissions in a many-to-one network, where channel probing can be realized by broadcasting pilot signals from the base station. Notably, [1, 77] study opportunistic ALOHA under a collision model, with a basic idea being that in every slot each user transmits with a probability based on its own channel condition. While recent work [47] does not assume a base station in a wireless LAN, the transmitter node still needs to collect channel information from potential receivers, thereby serving as a tentative "virtual" base station. A key observation is that in the existing work on rate adaptation for ad hoc communications (see, e.g., [44, 76, 81]), a link continues transmission after a successful channel contention, no matter whether the channel condition is good or poor. Clearly, this leaves much room for improvement by devising channel-aware scheduling.

Unfortunately, little work has been done on developing channel-aware distributed scheduling to reap rich diversity gains for enhancing ad hoc communications. This is perhaps due to the fact that channel-aware distributed scheduling is indeed challenging, since *the distributed nature of ad hoc communications dictates that each link has no knowledge of others' channel conditions* (in fact, even its own channel condition is unknown before channel probing). A principal goal of the study in this chapter is to fill this void, and obtain a rigorous understanding of distributed opportunistic scheduling (DOS) for ad-hoc communications.

Specifically, we consider a single-hop ad-hoc network where all links can hear others' transmissions. In such a network, links contend for the same channel using random access, and a collision model is assumed which indicates that at most one link can transmit successfully at each time. We assume that after a successful contention, the channel condition of the successful link is measured (e.g., by using some pilot signals embedded in the probing packets). Due to channel fading, the link condition corresponding to this successful channel probing can be either good or poor. In the latter case, data packets have to be transmitted at low rates, leading to possible throughput degradation. A plausible alternative is to let this link give up this transmission opportunity, and allow all the links re-contend for the channel, in the hope that some link with a better channel condition can transmit after the re-contention. Intuitively speaking, because different links at different time slots experience different channel conditions, it is likely that after further probing, the channel can be taken by a link with a better channel condition, resulting in possible higher throughput. In this way, the multiuser diversity across links and the time diversity across slots can be exploited in an opportunistic manner. It is in this sense that we call this process of joint probing and scheduling "distributed opportunistic scheduling" [101]. We should caution that on the other hand, each channel probing comes with a cost in terms of the contention time, which could be used for data transmission.

Clearly, there is a *tradeoff* between the throughput gain from better channel conditions and the cost for further channel probing. The desired tradeoff boils down to judiciously choosing the optimal stopping rule for channel probing, in order to maximize the throughput. In this chapter, we obtain a systematic characterization of this tradeoff by appealing to optimal stopping theory [35, 39, 83], and explore channel-aware distributed scheduling to exploit multiuser diversity and time diversity for wireless ad-hoc networks in an opportunistic manner.

2.1.2. Summary of Main Results

We study distributed opportunistic scheduling for network throughput optimization. We start with the basic case where all links have the same channel statistics. Recall that when a link discovers that its channel condition is relatively poor after a successful channel contention, it can skip the transmission opportunity so that some link with a better condition would have the chance to transmit in the next round channel probing. We should point out that there is no guarantee for this to happen due to the stochastic nature of random contention and time varying channel conditions. Nevertheless, as channel probing continues, the likelihood of reaching a better channel condition increases. In a nutshell, distributed opportunistic scheduling boils down to a process of joint channel probing and scheduling.

Mathematically speaking, building on optimal stopping theory [35, 39, 83], we cast the problem as a *maximal rate of return* problem, where the rate of return refers to the average throughput. As noted above, since the cost, in terms of the contention duration, is random, we use the Maximal Inequality to establish the existence of the optimal stopping rule. Then, we develop the optimal strategy for distributed opportunistic scheduling, by characterizing the optimal stopping rule to "control" the channel probing process and hence to maximize the overall throughput. In particular, we show that the optimal strategy is a *pure threshold policy*, [1] in the sense that the decision on further channel probing or data transmission is based on the local channel condition only, and the threshold is invariant in time. Therefore, it is amenable to distributed implementation. Furthermore, it turns out that the optimal threshold can be chosen to be the maximum network through-

---

[1] A threshold policy is called pure if the threshold is invariant in time.

put, which can be obtained by solving a fixed point equation. We then generalize the above study to the case with heterogeneous links, where different links may have different channel statistics. Due to the channel heterogeneity, the channel conditions corresponding to consecutive successful channel probings may follow different distributions. Again, we show that the optimal strategy for joint channel probing and distributed scheduling is a pure threshold policy. Somewhat surprisingly, the optimal thresholds turn out to be the same across all the links regardless of the channel statistics and contention probabilities. We further devise iterative algorithms to compute the optimal threshold. We also generalize the studies to take into account fairness requirements.

2.1.3. Related Work and Organization

As noted above, there has been much work on centralized opportunistic scheduling (e.g., [3], [6], [24], [25], [54], [59], [60], [82], [93]), channel-aware ALOHA (e.g., [1, 77]) and MAC design with rate adaptation (e.g., [44, 49, 76, 81]). Most relevant to our study are perhaps (e.g., [1, 44, 77, 81]). The main differences between this study and the studies [1, 77] lie in the following two aspects: 1) We consider ad hoc communications assuming no centralized coordination, and the transmission scheduling is done distributively; and 2) the transmitter nodes have no knowledge of other links' channel conditions, and even their own channel conditions are not available before contention. These limitations, dictated by the distributed nature of ad hoc communications, pose great challenges for exploiting channel diversity in distributed scheduling. A major difference between our study and the studies in [44, 81] is that our scheme allows links to opportunistically utilize the channel whereas in the schemes in [44, 81] the transmission rate is adapted based on the current channel condition, regardless of whether the channel condition is poor or good.

Along a different avenue, opportunistic channel probing for single-user multichannel systems has been studied in [41], where the basic idea is to opportunistically probe and select a transmission channel among multiple channels between the transmitter node and the receiver node. In contrast, in this study, we consider multiple links (each with its own transmitter and receiver) sharing one single channel and explore distributed scheduling, assuming that each link has no knowledge of other links' channel conditions.

The rest of the chapter is organized as follows. In Section 2.2, we introduce some background knowledge on optimal stopping theory. Section 2.3 presents the model for random-access based channel probing and scheduling. In Section 2.4, we formulate the problem of joint channel probing and scheduling from the network-centric perspective. We further characterize the optimal stopping rule under fairness constraints in Section 2.5, and time constraints in Section 2.6. In Section 2.7, we provide numerical examples to corroborate the theoretic results. Finally, Section 2.8 concludes the paper.

2.2. A Preliminary on Optimal Stopping Theory

As noted above, in an ad-hoc network with many links, when a link discovers that its channel condition is "relatively poor" after a successful channel contention, it can either transmit or skip this opportunity so that in the next round some link with a better condition would have the chance to transmit. This is intimately related to the optimal stopping strategy in sequential analysis [39].

Simply put, an optimal stopping rule is a strategy for deciding when to take a given action based on the past events in order to maximize the average return, where

18

the return is the net gain (the difference between the reward and the cost) [35, 39, 83]. More specifically, let $\{Z_1, Z_2, \ldots\}$ denote a sequence of random variables, and $\{y_0, y_1(z_1), y_2(z_1, z_2), \ldots, y_\infty(z_1, z_2, \ldots)\}$ a sequence of real-valued reward functions. The reward is $y_n(z_1, \ldots, z_n)$ if the strategy chooses to stop at time n. The theory of optimal stopping is concerned with determining the stopping time N to maximize the expected reward $E[Y_N]$; and in general N is called a *stopping time* if $\{N = n\} \in \mathcal{F}_n$, where $\mathcal{F}_n$ is the σ–algebra generated by $\{Z_j, j \leq n\}$.

2.3. System Model

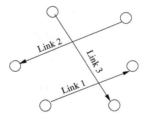

Figure 2.1. An example of a single-hop ad-hoc network.

Random access is widely used for medium access control in wireless ad hoc networks. Consider a single-hop ad-hoc network with M links (see Fig. 2.1), where link m contends for the channel with probability p_m, $m = 1, \ldots, M$. A collision model is assumed for the random access, where a channel contention of a link is said to be successful if no other links transmit at the same time. We assume that the local channel condition can be obtained after a successful channel contention. Accordingly, the overall successful channel probing probability in each slot, p_s, is then given by $\sum_{m=1}^{M}(p_m \prod_{i \neq m}(1 - p_i))$ [77]. (To avoid triviality, we assume that $p_s > 0$.)

For convenience, we call the random duration of achieving a successful channel contention as one round of channel probing. It is clear that the number of slots (denoted by K) for a successful channel contention (probing) is a Geometric random variable, i.e., $K \sim Geometric(p_s)$. Let τ denote the duration of mini-slot for channel contention. It follows that the random duration corresponding to one round of channel probing is $K\tau$, with expectation τ/p_s.

Let $s(n)$ denote the successful link in the n-th round of channel probing, and $R_{n,s(n)}$ denote the corresponding transmission rate. In wireless communications, $R_{n,s(n)}$ depends on the time varying channel condition, and hence is random. Following the standard assumption on the block fading channel in wireless communications [44, 81], we assume that the rate $R_{n,s(n)}$ remains constant for a duration of T, where T is the data transmission duration and is no greater than the channel coherence time. [2]

To get a more concrete sense of joint channel probing and distributed scheduling, we depict in Fig. 2.2 an example with N rounds of channel probing and one single data transmission. Specifically, suppose after the first round of channel probing with a duration of $K_1\tau$, the rate of link $s(1)$, $R_{1,s(1)}$, is small (indicating a poor channel condition); and as a result, $s(1)$ gives up this transmission opportunity and let all the links re-contend. Then, after the second round of channel probing with a duration of $K_2\tau$, link $s(2)$ also gives up the transmission because $R_{2,s(2)}$ is small. This continues for N rounds until link $s(N)$ transmits because $R_{N,s(N)}$ is good.

In this study, we provide a systematic study on distributed opportunistic scheduling by using optimal stopping theory. We first impose the following assumption on the

---

[2]Channel coherence time refers to the time during which the channel condition remains more or less unchanged.

transmission rates across different rounds of channel probing.

A1) $\{R_{n,s(n)}, n = 1, 2, \ldots\}$ are independent.

We note that the above condition holds in many practical scenarios of interest, and the rational behind is as follows: 1) in a multi-user wireless network, the likelihood of one link (say link m) achieving two consecutive successful channel probing, $p_m^2 \prod_{i \neq m}(1 - p_i)^2$, is fairly small, especially when the number of links is large; and 2), even if this happens, it is reasonable to assume that the channel conditions corresponding to two successful channel probings are independent since the channel probing duration in between is comparable to the channel coherence time.

2.4. Distributed Opportunistic Scheduling For Network Throughput Optimization

We study distributed opportunistic scheduling, namely, joint channel probing and distributed scheduling, to maximize the overall network throughput. In particular, building on optimal stopping theory, we cast the problem as *maximizing the rate of return*, where the rate of return refers to the average throughput [39]. For convenience, let $R_{(n)}$ denote the rate corresponding to the n-th round successful channel probing, i.e., $R_{(n)} = R_{n,s(n)}$. Without loss of generality, we assume that the second moment of $R_{(n)}$ exists.

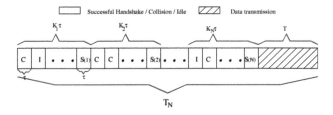

Figure 2.2. A sample realization of channel probing and data transmission

As illustrated in Fig. 2.2, after one round of channel probing, a stopping rule N decides whether the successful link carries out data transmission, or simply skips this opportunity and let all the links re-contend. Suppose that this game on joint channel probing and transmission is carried out L times, and let $\{N_1, N_2, \ldots, N_L\}$ denote the corresponding stopping times. Let T_{N_l} denote the l-th realization of the duration for probing and data transmission. Then, appealing to the Renewal Theorem, we have that

$$x_L = \frac{\sum_{l=1}^{L} R_{(N_l)} T}{\sum_{l=1}^{L} T_{N_l}} \longrightarrow \frac{E[R_{(N)} T]}{E[T_N]} \quad a.s., \tag{2.1}$$

where $E[R_{(N)} T]/E[T_N]$ is the rate of return [39]. Clearly, $R_{(N)}$ and T_N are stopped random variables since N is a stopping time. Accordingly, the distributions of $R_{(N)}$ and T_N depend on that of the stopping time N. Define

$$Q \triangleq \{N : N \geq 1, E[T_N] < \infty\}. \tag{2.2}$$

It then follows that the problem of maximizing the long-term average throughput can be cast as a maximal-rate-of-return problem, in which a key step is to characterize the optimal stopping rule N^* and the optimal throughput x^*, as

$$N^* \triangleq \arg\max_{N \in Q} \frac{E[R_{(N)} T]}{E[T_N]}, \quad x^* \triangleq \sup_{N \in Q} \frac{E[R_{(N)} T]}{E[T_N]}. \tag{2.3}$$

2.4.1. Optimal Stopping Rule for Throughput Maximization

We now exploit optimal stopping theory [35, 39, 83] to solve the problem in (2.3).

2.4.1.1. The Case with Homogeneous Links

For ease of exposition, we first consider a network with homogenous links where all links have the same channel statistics with the same distribution $F_R(r)$. By **A1**, $\{R_{(n)}, n = 1, 2, \ldots\}$ is a sequence of i.i.d. random variables with distribution $F_R(r)$.

Observe that different from standard optimal stopping problems, the cost in terms of the probing duration is random due to the stochastic nature of channel probing. In light of this, we use the Maximal Inequality to establish the existence of the optimal stopping rule. We have the following proposition.

Proposition 2.4.1. *a) The optimal stopping rule N^* for distributed opportunistic scheduling exists, and is given by*

$$N^* = \min\{n \geq 1 : R_{(n)} \geq x^*\}. \tag{2.4}$$

b) The maximum throughput x^* is an optimal threshold, and is the unique solution to

$$E(R_{(n)} - x)^+ = \frac{x\tau}{p_s T}. \tag{2.5}$$

Proof: The proof hinges heavily on the tools in optimal stopping theory [39]. More specifically, based on Theorem 1 in [39, Chapter 6], in order to maximize the average throughput $\frac{E[R_{(N)}T]}{E[T_N]}$, a key step is to find an optimal stopping algorithm $N(x)$ such that

$$V^*(x) = E[R_{(N(x))}T - xT_{N(x)}] = \sup_{N \in Q} E[R_{(N)}T - xT_N].$$

It then follows from Theorem 1 in [39, Chapter 3] that $N(x)$ exists if the following conditions are satisfied:

$$E \sup_n Z_n < \infty, \quad \text{and} \quad \limsup_{n \to \infty} Z_n = -\infty \ a.s., \tag{2.6}$$

where $Z_n \triangleq R_{(n)}T - xT_n$, $T_n \triangleq \sum_{j=1}^n K_j \tau + T$, and $K_j, j = 1, 2, \ldots, n$, denote the number of contentions during the jth channel probing.

The rest of the proof has two main steps. Step 1: we establish the existence of the optimal stopping rule $N(x)$; Step 2: we characterize the optimal strategy N^*.

Step 1: It is clear that $\limsup_{n\to\infty} Z_n \longrightarrow -\infty$.

Observe that $E[\sup_n Z_n]$ is upper-bounded by

$$E[\sup_n Z_n]$$
$$\leq E\left[\sup_n \left\{ R_{(n)}T - n x \tau \left(\frac{1}{p_s} - \epsilon\right)\right\}\right] - Tx + E\left[\sup_n \sum_{j=1}^n x\tau \left(\frac{1}{p_s} - \epsilon - K_j\right)\right] \quad (2.7)$$

where ϵ is chosen such that $0 < \epsilon < 1/p_s$. It then follows from the Maximal Inequalities

in Theorem 1 and Theorem 2 in [39, Chapter 4] that the first term and the last term of

the right hand side of (2.7) are both finite, and hence $E[\sup_n Z_n] < \infty$.

Step 2: Next, we characterize $N(x)$ and N^*. It can be shown that the optimal stopping

rule $N(x)$ is given by

$$N(x) = \min\{n \geq 1 : R_{(n)}T \geq V^*(x) + xT\}, \quad (2.8)$$

and $V^*(x)$ satisfies the following *optimality equation*:

$$E[\max(R_{(n)}T - xT - Kx\tau, V^*(x) - Kx\tau)] = V^*(x). \quad (2.9)$$

Note that $V^*(x^*) = 0$ from Theorem 1 in [39, Chapter 6] and (2.9) becomes $E[R_{(n)}-x^*]^+ = \frac{x^*\tau}{p_s T}$ since $E[K] = 1/p_s$. The optimal stopping rule (2.8) now becomes $N^* = \min\{n \geq 1 : R_{(n)} \geq x^*\}$.

Next we show that (2.5) has a unique solution. We first note that $f(x) \triangleq E[R_{(n)} - x]^+$

is continuous in x. To see this, let $\{x_l, l = 1, 2, \ldots\}$ be a sequence of real positive numbers,

and $\lim_{l\to\infty} x_l = x$, then $R_{(n)} - x_l \to R_{(n)} - x$ almost surely. Since $|R_{(n)} - x_l| \leq R_{(n)}$, we have

that $f(x_l) \to f(x)$ by using Dominated Convergence Theorem [35]. Since $f(x)$ decreases

from $E[R_{(n)}]$ to 0 and the right hand side of (2.5) strictly increases from 0 to ∞ as x

grows, it follows that (2.5) has a unique finite solution. ∎

Remarks: 1) Proposition 2.4.1 reveals that the optimal stopping rule N^* for distributed opportunistic scheduling is a pure threshold policy, and the stopping decision can be made based on the current rate only. Accordingly, the optimal channel probing and scheduling strategy takes the following simple form: If the successful link discovers that the current rate $R_{(n)}$ is higher than the threshold x^*, it transmits the data with rate $R_{(n)}$; otherwise, it skips this transmission opportunity (e.g., by skipping CTS), and then the links re-contend.

2) We note that the maximum throughput x^* is unique, but the optimal threshold in (2.4) may not be unique in general. It is not difficult to show the uniqueness of the optimal threshold in the continuous rate case with $f(r) > 0$, $\forall \, r > 0$. In contrast, in the discrete rate case, changing the threshold in between two adjacent quantization levels would not affect its optimality since the new threshold policy achieves the same throughput. (In what follows, for the discrete rate case, we treat the thresholds in between two adjacent quantization levels "effectively" the same.)

3) It can be shown that

$$E[T_N] = \frac{\tau}{p_s} E[N] + T. \tag{2.10}$$

Based on (2.10) and the proof of Theorem 2.4.1, it can also be shown that if the random contention time $K\tau$ is replaced with a constant probing time τ/p_s, the optimal stopping rule (2.4) and the optimal throughput remain the same.

Based on the structure of the optimal stopping rule N^* in (2.4), we have the following corollary.

Corollary 2.4.1. *a) The stopping time N^* is geometrically distributed with parameter $1 - F_R(x^*)$.*

b) The stopped random variable R_{N^*} has the following distribution:

$$F_{R_{N^*}}(r) = \begin{cases} \frac{F_R(r)-F_R(x^*)}{1-F_R(x^*)}, & r \geq x^*; \\ \\ 0, & otherwise. \end{cases}$$

c) The stopped random variable $\frac{T_{N^*}-T}{\tau}$ is geometrically distributed with parameter $p_s[1 - F_R(x^*)]$.

Proof: a) and b) follows from the definition of N^* in (2.4) directly. To examine c), we first note that $\frac{T_{N^*}-T}{\tau} = \sum_{i=1}^{N^*} K_i$. It then can be shown that the moment generating function of $\sum_{i=1}^{N^*} K_i$ is $\frac{p_s(1-F_R(x^*))e^t}{1-(1-p_s(1-F_R(x^*)))e^t}$. Therefore, we conclude that $\sum_{i=1}^{N^*} K_i$ is geometrically distributed with parameter $p_s(1 - F_R(x^*))$. ∎

Part a) of Corollary 2.4.1 indicates that the channel probing process would stop in a finite time almost surely. It follows from part b) and c) of Corollary 2.4.1 that

$$\frac{E[R_{N^*}T]}{E[T_{N^*}]} = \frac{\int_{x^*}^{\infty} r\,dF_R(r)}{\frac{\delta}{p_s} + 1 - F_R(x^*)}, \tag{2.11}$$

where $\delta = \tau/T$.

We note that the maximum throughput x^* is obtained by solving the fixed point equation (2.5), which in general does not admit a closed-form solution. In what follows, we derive a lower bound and an upper bound on x^*. We have the following proposition.

Proposition 2.4.2.

$$x^L \leq x^* \leq x^U,$$

where x^L and x^U are given by

$$x^L \triangleq \frac{E[R]}{\frac{\delta}{p_s} + 1}, \qquad x^U \triangleq \sqrt{\frac{E[R^2]}{2\frac{\delta}{p_s}}}. \tag{2.12}$$

Remarks: 1) Observe that x^L is the throughput of the OAR scheme in [81], which can be viewed as a degenerated stopping algorithm with zero threshold.

2) Note that $\sqrt{\frac{E[R^2]}{2\frac{\delta}{p_s}}}$ is the maximum throughput corresponding to the optimal genie-aided scheduling when channel realizations are known *a priori*. Indeed, this can be seen from the proof of Proposition 2.4.2.

Proof of Proposition 2.4.2: It is clear that x^L is achieved by a special stopping algorithm (which stops at the very first time). Therefore, by the definition of x^*, $x^L \leq x^*$.

To show that x^U is an upper-bound on x^*, recall that from Remark 3) for Prop. 2.4.1, replacing the random contention period, $K\tau$, with a constant access time, τ/p_s, would yield the same optimal long-term average rate x^*. Accordingly, the upper-bound derived for the constant access time case also serves an upper-bound on x^*.

Observe that for any constant x,

$$E\left[\sup_n \left\{ R_{(n)}T - x \cdot \left(\frac{\tau}{p_s}n + T\right)\right\}\right] = E\left[\sup_n \left(R_{(n)}T - x\frac{\tau}{p_s}n\right)\right] - xT$$
$$\leq \frac{E[T^2 R^2]}{2\frac{x\tau}{p_s}} - xT, \tag{2.13}$$

where the last inequality is from Theorem 1 in [39, Chapter 4]. Plugging $x = \sqrt{\frac{E[R^2]}{2\frac{\delta}{p_s}}}$ into (2.13) yields that

$$E\left[\sup_n \left\{ R_{(n)}T - \sqrt{\frac{E[R^2]}{2\frac{\delta}{p_s}}} \cdot \left(\frac{\tau}{p_s}n + T\right)\right\}\right] \leq 0. \tag{2.14}$$

Furthermore, we have that

$$E\left[R_N^*T - x^* \cdot \left(\frac{\tau}{p_s}N^* + T\right)\right] = 0. \tag{2.15}$$

Combining (2.14) and (2.15), we have that

$$E\left[\sup_n \left\{R_nT - \sqrt{\frac{E[R^2]}{2\frac{\delta}{p_s}}} \cdot \left(\frac{\tau}{p_s}n + T\right)\right\}\right] \leq E\left[R_{N^*}T - x^* \cdot \left(\frac{\tau}{p_s}N^* + T\right)\right]$$

$$\stackrel{(a)}{=} \sup_{N \in Q} E\left[R_NT - x^* \cdot \left(\frac{\tau}{p_s}N + T\right)\right]$$

$$\stackrel{(b)}{\leq} E\left[\sup_n \left\{TR_n - x^* \cdot \left(\frac{\tau}{p_s}n + T\right)\right\}\right] \tag{2.16}$$

where (a) is by the definition of N^*, and (b) can be obtained using the same technique as in Fatou's Lemma [21].

It is clear that for any $x_1 \leq x_2$,

$$E\left[\sup_n \left\{R_{(n)}T - x_1 \cdot \left(\frac{\tau}{p_s}n + T\right)\right\}\right] \geq E\left[\sup_n \left\{R_{(n)}T - x_2 \cdot \left(\frac{\tau}{p_s}n + T\right)\right\}\right].$$

It follows from (2.16) that $x^* \leq \sqrt{\frac{E[R^2]}{2\frac{\delta}{p_s}}}$. ∎

2.4.1.2. The Case with Heterogeneous Links

In the above, it is assumed that all links have the same channel statistics. As a result, $R_{n,s(n)}$ follows the same distribution $F_R(r)$. In many practical scenarios, it is likely that different links may have different channel statistics. As a result, if $s(n+1) \neq s(n)$, $R_{n,s(n)}$ and $R_{n+1,s(n+1)}$ may follow different distributions. Nevertheless, we can treat $R_{n,s(n)}$ as a compound random variable. Accordingly, a key step is to characterize the distribution of $R_{n,s(n)}$ for the heterogeneous case.

To this end, let $F_m(\cdot)$ denote the distribution for each link $m \in \{1, 2, \ldots, M\}$. It can be shown that

$$P(R_{(n)} \leq r) = E\left[P(R_{n,m} \leq r)|s(n) = m\right] = \sum_{m=1}^{M} \frac{p_{s,m}}{p_s} F_m(r), \tag{2.17}$$

where $p_{s,m} \triangleq p_m \prod_{i \neq m}(1 - p_i)$ is the successful probing probability of user m. Based on (2.17), it is clear that $R_{(n)}$ is a compound random variable whose distribution is a "mixed" version of that across the links. We have the following proposition regarding the optimal threshold policy.

Proposition 2.4.3. *The maximum throughput x^* in the heterogeneous case is an optimal threshold, and is the unique solution to the following equation:*

$$x = \frac{\sum_{m=1}^{M} p_{s,m} \int_x^\infty r \, dF_m(r)}{\delta + \sum_{m=1}^{M} p_{s,m} \left(1 - F_m(x)\right)}. \tag{2.18}$$

Remarks: For the heterogeneous case, *a priori*, it is not clear that different links would have different thresholds or not since their channel statistics are different. However, Proposition 2.4.3 indicates that in the optimal strategy the threshold is the same for all the links (again, for the discrete rate case, we treat the thresholds in between two adjacent quantization levels "effectively" the same). Our intuition is as follows: When all the links have the same threshold, links with better channel conditions would have a higher likelihood to transmit accordingly.

2.4.2. Iterative Computation Algorithm for x^*

In the following, we devise an iterative algorithm to compute x^*. To this end, rewrite (2.18) as $x = \Phi(x)$, with

$$\Phi(x) \triangleq \frac{\sum_{m=1}^{M} p_{s,m} \int_x^\infty r \, dF_m(r)}{\delta + \sum_{m=1}^{M} p_{s,m} \left(1 - F_m(x)\right)}. \tag{2.19}$$

Accordingly, we propose the following iterative algorithm for computing x^*.

$$x_{k+1} = \Phi(x_k), \text{ for } k = 0, 1, 2, \ldots, \tag{2.20}$$

where x_0 is the initial value. We have the following proposition on the convergence of the above iterative algorithm.

Proposition 2.4.4. *The iterates generated by algorithm (2.20) converge to x^* for any positive initial value x_0.*

A standard approach for establishing the convergence of iterative fixed point algorithms is via the Contraction (or Pseudo-Contraction) Mapping Theorem [15], which is unfortunately not applicable here since $\Phi(x)$ is not a pseudo-contraction mapping in some cases. For instance, suppose for any m, $f_m(r)$ is given by

$$f_m(r) = \begin{cases} 0, & r < 0 \\ 0.01, & 0 \leq r < 96 \\ 0.005(r - 94), & 96 \leq r < 98 \\ 0.02(r - 97)^{-3}, & r \geq 98 \end{cases} \tag{2.21}$$

Let $p_{s,m} = 0.99/M$ and $\delta = 0.05$. The corresponding optimal point $x^* = 72.82$. However,

$$|\Phi(95.5) - x^*| = |45.88 - 72.82| > |95.5 - 72.82|,$$

which violates the condition for pseudo-contraction mapping.

In light of the above observation, we provide in the following a new proof for the convergence of iterative algorithm (2.20). We first need the following lemma.

Lemma 2.4.1. x^* *is a global maximum point of $\Phi(x)$.*

It can be shown that $\Phi(x)$ is the average network throughput under the following stopping rule:

$$N = \min\{n \geq 1 : R_{(n)} \geq x\}.$$

The proof then follows from Proposition 2.4.1.

Proof of Proposition 2.4.4: From Lemma 2.4.1 and Proposition 2.4.3, it is clear that $y = \Phi(x)$ only intersects $y = x$ at the point x^*. See Fig. 2.3 for a pictorial illustration. This, together with the fact that $\Phi(0) > 0$, yields that

$$\Phi(x) \geq x, \forall\, x \leq x^*; \Phi(x) \leq x, \forall\, x > x^*. \tag{2.22}$$

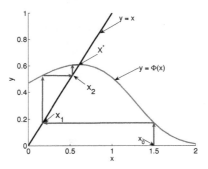

Figure 2.3. Convergence of the iterative algorithm (2.20)

Without loss of generality, we can assume that $x_0 \leq x^*$ (we note that if $x_0 > x^*$, $x_1 = \Phi(x_0) \leq \Phi(x^*) = x^*$ according to Lemma 2.4.1). Next, suppose that $x_k \leq x^*$. From (2.22), we obtain that $x_k \leq \Phi(x_k) = x_{k+1} \leq x^*$, where the last inequality is due to the fact that $\Phi(x_k) \leq \Phi(x^*) = x^*$ from Lemma 2.4.1. Since $0 < x_0 \leq x^*$, it follows that $\{x_k, k = 1, 2, \ldots\}$ is a monotonically increasing positive sequence with an upper-bound x^*. As a result, the sequence $\{x_k, k = 1, 2, \ldots\}$ converges to a limit, denoted as x_∞.

To show that $x_\infty = x^*$, we rewrite $x_{k+1} = \Phi(x_k)$ as

$$E[R_{(n)} - x_k]^+ - x_k\frac{\delta}{p_s} = (x_{k+1} - x_k)\left(\frac{\delta}{p_s} + \sum_{m=1}^M \frac{p_{s,m}}{p_s}\left(1 - F_m(x_k)\right)\right). \qquad (2.23)$$

Observe that $E[R_{(n)} - x]^+$ is continuous in x (see the proof of Proposition 2.4.1), $x_{k+1} -$

$x_k \to 0$ as $k \to \infty$, and $\frac{\delta}{p_s} + \sum_{m=1}^M \frac{p_{s,m}}{p_s}\left(1 - F_m(x_k)\right) \leq \frac{\delta}{p_s} + 1 < \infty$. Therefore, taking

limits on both sides of (2.23) yields that $E[R_{(n)} - x_\infty]^+ - \delta x_\infty/p_s = 0$.

Since from Proposition 2.4.3 $E[R_{(n)} - x]^+ = x\frac{\delta}{p_s}$ has a unique solution, we conclude

that $x_\infty = x^*$. ∎

2.5. Distributed Opportunistic Scheduling Under General Fairness Constraints

In the above studies, the optimal distributed scheduling is targeted at maximizing

the overall network throughput. We next generalize the study to take into account fair-

ness requirements. Under fairness constraints, the objective of distributed opportunistic

scheduling boils down to maximizing the total network utility function, where user m's

utility is a function of its rate and serves as a measure of satisfaction that user m has

from sharing the channel. For example, the reward function (utility function), denoted

$\{U_m(r), \forall\, m\}$ can take the following form [68]:

$$U_{m,\kappa}(r) = \begin{cases} w_m \log r, & \text{if } \kappa = 1 \\ w_m(1 - \kappa)^{-1}r^{1-\kappa}, & \kappa \geq 0, \kappa \neq 1. \end{cases} \qquad (2.24)$$

Then, the optimal strategy for distributed opportunistic scheduling is to characterize

the optimal stopping rule N_U^* for maximizing the return rate of the total network utility,

i.e.,

$$N_U^* \triangleq \arg\max_{N \in Q} \frac{E[U(R_N)T]}{E[T_N]}, \qquad (2.25)$$

Interestingly, when $\kappa = 0$, the above problem degenerates to the problem of maximizing the overall throughput. Furthermore, when $\kappa = 1$, *proportional fairness* is achieved by N_U^*.

It is not difficult to see that the optimal stopping rule N_U^* can be derived, along the same line as in Proposition 2.4.1 and 2.4.3. We note that this study can be further extended to incorporate more complicated fairness constraints.

2.6. Distributed Opportunistic Scheduling Under Time Constraints

In this section, we generalize the above study to the case with time constraints. Specifically, recall that in (2.2), $E[T_N] < \infty$. We now impose an additional constraint with $E[T_N] \leq \alpha$ for some $\alpha > 0$. This is useful for time-sensitive applications with delay constraints, e.g., streaming traffic.

We have the following problem:

$$\underset{N \in Q_\alpha}{\text{maximize}} \ \frac{E[R_{(N)}T]}{E[T_N]}, \tag{2.26}$$

where

$$Q_\alpha \triangleq \{N : N \geq 1, E[T_N] \leq \alpha\}. \tag{2.27}$$

Comparing Q_α with Q, we see that the new constraint dictates that the average duration of probing and transmission must be no greater than α.

To solve (2.26), we first need the following lemma.

Lemma 2.6.2. *Let $\{X_n, n = 1, 2, \ldots\}$ be a sequence of i.i.d. random variables. Suppose there are two sequences $\{Y_n, n = 1, 2, \ldots\}$ and $\{T_n, n = 1, 2, \ldots\}$ both $\mathcal{F}_n$-measurable, where $\mathcal{F}_n$ is the σ-field generated by $\{X_j, j \leq n\}$. Assume that $0 < T_1 \leq T_2 \leq \ldots$ a.s.. Define $Z_n \triangleq Y_n - xT_n$, where x is a positive constant.*

If there exists a constant x^* *such that* $\sup_{N \in Q_\alpha} E[Y_N - x^*T_N] = 0$, *then* $\sup_{N \in Q_\alpha} E[Y_N]/E[T_N] = x^*$. *Moreover, if* $N^* \in Q_\alpha$ *is a stopping algorithm such that* $E[Y_{N^*} - x^*T_{N^*}] = \sup_{N \in Q_\alpha} E[Y_N - x^*T_N] = 0$, *then* N^* *is optimal in the sense that* $E[Y_N^*]/E[T_N^*] = \sup_{N \in Q_\alpha} E[Y_N]/E[T_N]$.

The proof of the above lemma follows directly from Theorem 1 in [39, Chapter 6]. For convenience, define

$$N_\alpha^* = \arg\max_{N \in Q_\alpha} \frac{E[R_{(N)}T]}{E[T_N]}, \quad x_\alpha^* \triangleq \sup_{N \in Q_\alpha} \frac{E[R_{(N)}T]}{E[T_N]}. \tag{2.28}$$

Using Lemma 2.6.2 and Theorems 4 and 5 in [53], we next derive the optimal scheduling policy N_α^* and the optimal throughput x_α^* in four steps:

Step 1: Based on Lemma 2.6.2, the main problem (2.26) is first transformed into the following problem:

$$\text{maximize } E[R_{(N)}T] - xE[T_N], \text{ subject to } N \in Q_\alpha. \tag{2.29}$$

Step 2: Next, consider the problem (2.29) with Lagrangian relaxation:

$$\text{maximize } E[R_{(N)}T] - (x + \lambda)E[T_N] + \lambda\alpha, \text{ subject to } N \in Q. \tag{2.30}$$

Define

$$N^*(x, \lambda) \triangleq \arg\max_{N \in Q} E[R_{(N)}T] - (x + \lambda)E[T_N], \tag{2.31}$$

and

$$V^*(x, \lambda) \triangleq E[R_{(N^*(x,\lambda))}T] - (x + \lambda)E[T_{N^*(x,\lambda)}]. \tag{2.32}$$

Following the same procedure as in the proof of Proposition 2.4.1, it can be shown that

$$N^*(x, \lambda) = \min\{n \geq 1 : R_{(n)} \geq x + \lambda + V^*(x, \lambda)/T\}, \tag{2.33}$$

and

$$E\left[\left(R_{(n)}T - (x + \lambda)T - V^*(x, \lambda)\right)^+\right] = \frac{(x + \lambda)\tau}{p_s}. \tag{2.34}$$

Consequently, by Corollary 2.4.1, we have that

$$E[T_{N^*(x,\lambda)}] = \frac{\tau}{p_s[1 - F_R((x + \lambda)T + V^*(x, \lambda))]} + T. \tag{2.35}$$

Step 3: Solve the dual problem:

$$\text{minimize } L_x(\lambda) = V^*(x, \lambda) + \lambda\alpha, \text{ subject to } \lambda \geq 0. \tag{2.36}$$

Let $\lambda^*(x) \triangleq \arg\min_{\lambda \geq 0} L_x(\lambda)$. By the complementary slackness condition in Theorem 4 [53] and (2.35), we have that

$$\lambda^*(x)\left[\frac{\tau}{p_s[1 - F_R((x + \lambda^*(x))T + V^*(x, \lambda^*(x)))]} + T - \alpha\right] = 0. \tag{2.37}$$

Step 4: Characterize x_α^* by solving the following equation:

$$L_x(\lambda^*(x)) = 0. \tag{2.38}$$

To this end, we consider the following two cases.

1. If $\lambda^*(x_\alpha^*) > 0$, then it follows from (2.37) that

$$(x_\alpha^* + \lambda^*(x_\alpha^*))T + V^*(x_\alpha^*, \lambda^*(x_\alpha^*)) = F_R^{-1}\left(1 - \frac{\tau}{p_s(\alpha - T)}\right). \tag{2.39}$$

To guarantee the existence of a solution to (2.39), we need the following condition on α.

A2) $\alpha \geq \frac{\tau}{p_s} + T$.

Intuitively speaking, Condition **A2)** is sensible because on average it takes a duration of $\frac{\tau}{p_s} + T$ for the network to transmit a data packet successfully.

Combining (2.39) and (2.34) yields that

$$E\left[\left(R_{(n)}T - F_R^{-1}\left(1 - \frac{\tau}{p_s(\alpha - T)}\right)\right)^+\right] = \frac{(x_\alpha^* + \lambda^*(x_\alpha^*))\tau}{p_s}. \qquad (2.40)$$

Using (2.40) in (2.39), we have that

$$\begin{aligned}V^*(x_\alpha^*, \lambda^*(x_\alpha^*)) &= F_R^{-1}\left(1 - \frac{\tau}{p_s(\alpha - T)}\right) \\ &\quad - \frac{p_s T}{\tau}E\left[\left(R_{(n)}T - F_R^{-1}\left(1 - \frac{\tau}{p_s(\alpha - T)}\right)\right)^+\right] \end{aligned} \qquad (2.41)$$

It then follows from (2.40) and (2.41)

$$\begin{aligned}&L_x(\lambda^*(x_\alpha^*)) \\ &= V^*(x_\alpha^*, \lambda^*(x_\alpha^*)) + \lambda^*(x_\alpha^*)\alpha \\ &= F_R^{-1}\left(1 - \frac{\tau}{p_s(\alpha - T)}\right) - \frac{p_s T}{\tau}E\left[\left(R_{(n)}T - F_R^{-1}\left(1 - \frac{\tau}{p_s(\alpha - T)}\right)\right)^+\right] \\ &\quad + \alpha\left[\frac{p_s}{\tau}E\left[\left(R_{(n)}T - F_R^{-1}\left(1 - \frac{\tau}{p_s(\alpha - T)}\right)\right)^+\right] - x_\alpha^*\right] \end{aligned} \qquad (2.42)$$

Since $L_x(\lambda^*(x)) = 0$, we have that

$$\begin{aligned}x_\alpha^* &= \frac{1}{\alpha}F_R^{-1}\left(1 - \frac{\tau}{p_s(\alpha - T)}\right) \\ &\quad + \frac{p_s}{\tau}\left(1 - \frac{T}{\alpha}\right)E\left[\left(R_{(n)}T - F_R^{-1}\left(1 - \frac{\tau}{p_s(\alpha - T)}\right)\right)^+\right] \\ &= \frac{p_s}{\tau}\left(1 - \frac{T}{\alpha}\right)E\left[\max\left(R_{(n)}T, F_R^{-1}\left(1 - \frac{\tau}{p_s(\alpha - T)}\right)\right)\right]. \end{aligned} \qquad (2.43)$$

Note that $\lambda^*(x_\alpha^*) > 0$ according to the assumption, i.e.,

$$\lambda^*(x_\alpha^*) = \frac{p_s}{\tau}E\left[\left(R_{(n)}T - F_R^{-1}\left(1 - \frac{\tau}{p_s(\alpha - T)}\right)\right)^+\right] - x_\alpha^* > 0. \qquad (2.44)$$

From (2.43), (2.44) is equivalent to

$$F_R^{-1}\left(1 - \frac{\tau}{p_s(\alpha - T)}\right) < \frac{p_s T}{\tau}E\left[\left(R_{(n)}T - F_R^{-1}\left(1 - \frac{\tau}{p_s(\alpha - T)}\right)\right)^+\right].$$

$$(2.45)$$

2. Otherwise, if $\lambda^*(x_\alpha^*) = 0$, we have that

$$L_x(\lambda(x_\alpha^*)) = V^*(x_\alpha^*, 0) = 0. \tag{2.46}$$

It follows from (2.34) that

$$E\left[\left(R_{(n)}T - x_\alpha^* T\right)^+\right] = \frac{x_\alpha^* \tau}{p_s}, \tag{2.47}$$

which is exactly (2.5). Therefore, $x_\alpha^* = x^*$.

Summarizing, we have the following proposition regarding the optimal scheduling policy for the constrained case.

Proposition 2.6.5. *Assume Condition A2. Let α^* denote the unique solution to the following equation (in x):*

$$F_R^{-1}\left(1 - \frac{\tau}{p_s(x-T)}\right) = \frac{p_s T}{\tau} E\left[\left(R_{(n)}T - F_R^{-1}\left(1 - \frac{\tau}{p_s(x-T)}\right)\right)^+\right]. \tag{2.48}$$

Case 1), when $\alpha \geq \alpha^*$, we have that $N_\alpha^* = N^*$ and $x_\alpha^* = x^*$;

Case 2), when $\alpha < \alpha^*$, then the optimal scheduling policy N_α^* is given by

$$N_\alpha^* = \min\left\{n \geq 1 : R_{(n)} \geq F_R^{-1}\left(1 - \frac{\tau}{p_s(\alpha - T)}\right)\right\}, \tag{2.49}$$

and the optimal throughput x_α^* is

$$x_\alpha^* = \frac{p_s}{\tau}\left(1 - \frac{T}{\alpha}\right) E\left[\max\left(R_{(n)}T, F_R^{-1}\left(1 - \frac{\tau}{p_s(\alpha - T)}\right)\right)\right].$$

Remarks: Note that the left side of (2.48) increases from 0 to ∞ as x grows from $\frac{\tau}{p_s} + T$ to ∞, and the right side of (2.48) strictly decreases from $\frac{p_s T}{\tau} E\left[R_{(n)}T\right]$ to 0. Therefore, α^* is well-defined. According to Proposition 2.6.5, α^* can be deemed as a critical time constraint: if the time constraint α is less than α^*, then the optimal scheduling policy has to take into account the imposed time constraint; if α is bigger than α^*, then the imposed time constraint does not affect the optimal scheduling policy at all.

2.7. Numerical Results

Needless to say, a key performance metric is the throughput gain of distributed opportunistic scheduling over the approaches without using optimal stopping. For convenience, define the throughput gain as

$$g \triangleq \frac{x^* - x^L}{x^L}. \tag{2.50}$$

where x^L is the average throughput of the OAR scheme [81] without using optimal stopping, and $x^L = \Phi(0)$.

We consider the following two cases: 1) the continuous rate case based on Shannon capacity, and 2) the discrete rate case based on IEEE 802.11b.

2.7.1. Example 1: The Continuous Rate Case for Homogeneous Networks

Consider the case that the transmission rate is given by the Shannon channel capacity:

$$R(h) = \log(1 + \rho h) \text{ nats/s/Hz},$$

where ρ is the normalized average SNR, and h is the random channel gain corresponding to Rayleigh fading. It follows from (2.5) that

$$x^* = \Phi(\rho, x^*) = \frac{x^* \exp\left(-\frac{\exp(x^*)}{\rho}\right) + E_1\left(\exp(x^*)/\rho\right)}{\frac{\exp(-1/\rho)\delta}{p_s} + \exp\left(-\frac{\exp(x^*)}{\rho}\right)}, \tag{2.51}$$

where $E_1(x)$ is the *exponential integral function* defined as

$$E_1(x) \triangleq \int_x^\infty \frac{\exp(-t)}{t} dt.$$

Note that (2.51) can be further simplified as

$$x^* = \frac{\exp\left(\frac{1}{\rho}\right) E_1\left(\frac{\exp(x^*)}{\rho}\right)}{\frac{\delta}{p_s}}. \tag{2.52}$$

We have the following results on the optimal throughput x^* and the throughput gain $g(\rho)$.

Proposition 2.7.6. *a) The optimal throughput x^* is an increasing function of the average SNR ρ.*

b) The throughput gain $g(\rho)$ is maximized when $\rho \longrightarrow 0$, and

$$g(\rho) \longrightarrow \left(1 + \frac{\delta}{p_s}\right) \left.\frac{dx^*(\rho)}{d\rho}\right|_{\rho=0} - 1, \quad as \ \rho \longrightarrow 0, \tag{2.53}$$

where $\left.\frac{dx^*(\rho)}{d\rho}\right|_{\rho=0}$ is the root of

$$x \exp(x) = \frac{p_s}{\delta}. \tag{2.54}$$

Proof: To show that $x^*(\rho)$ is strictly increasing, it suffices to show that $\frac{dx^*(\rho)}{d\rho} > 0$ for any $\rho > 0$. To this end, we fix any arbitrary $\rho_0 > 0$, and let x_0^* be the corresponding optimal rate. Differentiating $\Phi(\rho, x)$ in (2.51) with respect to ρ yields

$$\frac{\partial \Phi(\rho, x)}{\partial \rho} = \frac{\exp\left(-\frac{\exp(x)-1}{\rho}\right) \frac{\exp(x)-1}{\rho^2}[x - \Phi(\rho, x)]}{\frac{\delta}{p_s} + \exp\left(-\frac{\exp(x)-1}{\rho}\right)} + \frac{\frac{1}{\rho}\frac{\exp(x)-1}{\rho} - \frac{\exp(\frac{1}{\rho})}{\rho^2} E_1\left(\frac{\exp(x)}{\rho}\right)}{\frac{\delta}{p_s} + \exp\left(-\frac{\exp(x)-1}{\rho}\right)}. \tag{2.55}$$

Since $x_0^* = \Phi(\rho_0, x_0^*)$, the first term of the right hand side of (2.55) is 0. Rewrite the second term as

$$\frac{1}{\rho_0} \exp\left(-\frac{\exp(x_0^*) - 1}{\rho_0}\right) - \frac{\exp(\frac{1}{\rho_0})}{\rho_0^2} E_1\left(\frac{\exp(x_0^*)}{\rho_0}\right)$$

$$> \frac{1}{\rho_0^2} \exp\left(\frac{1}{\rho_0}\right) \left[\rho_0 \exp\left(-\frac{\exp(x_0^*)}{\rho_0}\right) - \int_{\frac{\exp(x_0^*)}{\rho_0}}^{\infty} \frac{\exp(-t)}{\frac{\exp(x_0^*)}{\rho_0}} dt\right]$$

$$= \frac{1}{\rho_0} \exp\left(-\frac{\exp(x_0^*) - 1}{\rho_0}\right) \left(1 - \frac{1}{\exp(x_0^*)}\right)$$

$$> 0$$

indicating that $\frac{\partial \Phi(\rho_0, x_0^*)}{\partial \rho} > 0$. Therefore, for any sufficiently small $\Delta\rho > 0$, we have that

$$\Phi(\rho_0 + \Delta\rho, x_0^*) > \Phi(\rho_0, x_0^*). \tag{2.56}$$

Let $x_0^* + \Delta x^*$ be the corresponding optimal rate for $\rho_0 + \Delta\rho$. Due to the property of $\Phi(x)$, we observe that

$$x_0^* + \Delta x^* = \Phi(\rho_0 + \Delta\rho, x_0^* + \Delta x^*) \geq \Phi(\rho_0 + \Delta\rho, x_0^*) > \Phi(\rho_0, x_0^*) = x_0^*.$$

Therefore, $\Delta x^* > 0$ for $\Delta\rho > 0$. Following the same procedure, we can also show that $\Delta x^* < 0$ as $\Delta\rho < 0$. It follows that $\left.\frac{dx^*(\rho)}{d\rho}\right|_{\rho=\rho_0} = \lim_{\Delta\rho \to 0} \frac{\Delta x^*}{\Delta\rho} > 0$.

To prove part b), we can show that $g(\rho)$ is a decreasing function of ρ, and $g(\rho) \longrightarrow 0$ as $\rho \longrightarrow \infty$.

To examine the extreme case when $\rho \longrightarrow 0$, write $g(\rho)$ as follows using (2.52):

$$g(\rho) = \left(1 + \frac{p_s}{\delta}\right) \frac{E_1\left(\frac{\exp(x^*)}{\rho}\right)}{E_1\left(\frac{1}{\rho}\right)} - 1. \tag{2.57}$$

Using L'Hospital's rule, we can show that

$$g(\rho) \longrightarrow \left(1 + \frac{p_s}{\delta}\right) \exp\left(-\left.\frac{dx^*(\rho)}{d\rho}\right|_{\rho=0}\right) - 1, \text{ as } \rho \longrightarrow 0. \tag{2.58}$$

Next, we characterize $\left.\frac{dx^*(\rho)}{d\rho}\right|_{\rho=0}$. Rewrite (2.52) as follows

$$\frac{\delta}{p_s} x^* \exp\left(-\frac{1}{\rho}\right) = E_1\left(\frac{\exp(x^*)}{\rho}\right). \tag{2.59}$$

Taking differentiation on both sides of (2.59) with respect to ρ and rearranging yield that

$$\frac{\delta}{p_s}\frac{dx^*}{d\rho}\exp(x^*)\rho + \frac{\delta}{p_s}\frac{x^*}{\rho}\exp(x^*) = \exp\left(-\frac{\exp(x^*)-1}{\rho}\right)\left(1 - \frac{dx^*}{d\rho}\rho\right)\exp(x^*). \tag{2.60}$$

Let $\rho \longrightarrow 0$ in (2.60). Using the facts that $x^*(\rho) \longrightarrow 0$, $\frac{x^*(\rho)}{\rho} \longrightarrow \frac{dx^*}{d\rho}$ and $\frac{\exp(x^*(\rho))-1}{\rho} \longrightarrow \left.\frac{dx^*(\rho)}{d\rho}\right|_{\rho=0}$ as $\rho \longrightarrow 0$, it follows that $\left.\frac{dx^*(\rho)}{d\rho}\right|_{\rho=0}$ is the root of $x\exp(x) = p_s/\delta$. The proposition then follows from (2.58). ∎

Remarks: Prop. 2.7.6 reveals that the maximum gain is achieved in the low SNR region. In the extreme case when $\rho \to 0$, the gain is determined by the system parameters δ and

40

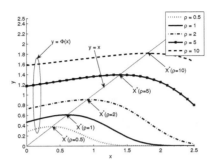

Figure 2.4. $\Phi(x)$ vs. x

p_s only. From (2.53) and (2.54), it is not difficult to see that the throughput gain increases as δ decreases or p_s increases. This is because a smaller δ or a larger p_s indicates that the probing cost is relatively insignificant.

We provide numerical examples to illustrate the above results. Unless otherwise specified, we assume that τ, T, p_s, and M are chosen such that $\delta = 0.1, p_s = \exp(-1)$.

Fig. 2.4 depicts $\Phi(\rho, x)$ as a function of x, for different ρ. It can be seen that the optimal average throughput x^* is strictly increasing in ρ. This can be further observed in Fig. 2.5, in which $x^*(\rho)$ is plotted as a function of ρ. In Table 2.1, we examine the convergence of the iterative algorithm (2.20). It can be seen that the convergence speed of the iterative algorithm (2.20) is fast, and the iterates approaches x^* usually within three or four iterations indifferent to the initial value x_0.

Table 2.2 illustrates that $g(\rho)$ is more significant in the low SNR region, and is a decreasing function of ρ. In Table 2.3, we present the maximum throughput gain $g(0)$ as a function of δ/p_s. It can be observed that $g(0)$ increases as the value of δ/p_s decreases. Intuitively speaking, a smaller value of δ indicates that the channel probing incurs less

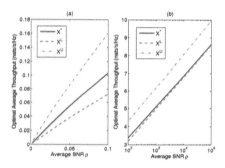

Figure 2.5. Optimal throughput x^* vs. average SNR ρ

Table 2.1. Convergence behavior of the iterative algorithm (2.20)

ρ	x_0	x_1	x_2	x_3	x^*
0.5	0.5	0.372213	0.384157	0.384283	0.384
1	0.5	0.603993	0.610418	0.610442	0.610
2	1.0	0.902320	0.906009	0.906014	0.906
5	1.0	1.357985	1.389121	1.389379	1.389
10	1.0	1.728041	1.807727	1.809031	1.809

overhead; and a larger value of p_s implies that the random access scheme yields higher throughput.

Table 2.2. Throughput gain

ρ	0.5	1	2	5	10
x^*	0.40	0.60	0.90	1.40	1.80
x^L	0.28	0.47	0.73	1.17	1.58
$g(\rho)$	42.8%	27.7%	23.3%	19.7%	13.9 %

2.7.2. Example 2: The Discrete Rate Case for Homogeneous Networks

Next, we study an example based on IEEE 802.11b, in which the transmission rates can be 2Mbps, 5.5Mbps and 11Mbps, with the following distribution:

$$R(h) = \begin{cases} 2 & \text{w.p. } p_2 = \frac{P(\gamma_2 \leq \rho h < \gamma_{5.5})}{P(\rho h \geq \gamma_2)} \\ 5.5 & \text{w.p. } p_{5.5} = \frac{P(\gamma_{5.5} \leq \rho h < \gamma_{11})}{P(\rho h \geq \gamma_2)} \\ 11 & \text{w.p. } p_{11} = \frac{P(\gamma_{11} \leq \rho h)}{P(\rho h \geq \gamma_2)}, \end{cases} \tag{2.61}$$

where $\gamma_2, \gamma_{5.5}$ and γ_{11} are the minimum SNR thresholds to support transmission rates of 2Mbps, 5.5Mbps and 11Mbps respectively.

Needless to say, the optimal throughput can be computed by using the general algorithm presented in (2.20). However, since the number of quantization levels is small (3 in

Table 2.3. Maximum throughput gain

δ/p_s	0.136	0.271	0.544	1.359	2.718
g (numerical)	76.4%	47.0%	25.7%	9.2%	3.5%
g (by (2.53))	76.6%	47.2%	25.7%	9.2%	3.5%

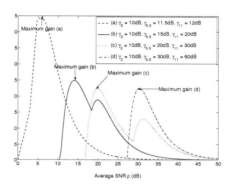

Figure 2.6. Throughput gain $g(\rho)$ as a function of average SNR ρ

this case), we can use "trial and error" to obtain the optimal throughput x^* as:

$$
\begin{aligned}
x^*(\rho) &= x^L \mathbf{I}\left(x^L < 2\right) + \frac{5.5p_{5.5} + 11p_{11}}{\frac{\delta}{p_s} + 1 - p_2} \mathbf{I}\left(2 \le \frac{5.5p_{5.5} + 11p_{11}}{\frac{\delta}{p_s} + 1 - p_2} < 5.5\right) \\
&+ \frac{11p_{11}}{\frac{\delta}{p_s} + p_{11}} \mathbf{I}\left(5.5 \le \frac{11p_{11}}{\frac{\delta}{p_s} + p_{11}} < 11\right),
\end{aligned}
\tag{2.62}
$$

where x^L is given by Proposition 2.4.2, $p_2, p_{5.5}$ and p_{11} can be computed from (2.61), and $\mathbf{I}(\cdot)$ is the indicator function.

As in centralized opportunistic scheduling, significant multiuser diversity gain can be achieved if the rate exhibits sufficient variation. Indeed, this can be observed in Fig. 2.6, where we plot the throughput gain of distributed opportunistic scheduling for different sets of thresholds $\{\gamma_2, \gamma_{5.5}, \gamma_{11}\}$. We also plot in Fig. 2.7 the probability of transmission rates as a function of ρ. It is clear that the "peaks" occur when the channel exhibits sufficient variations.

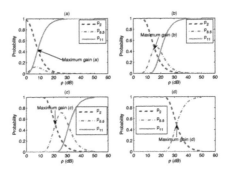

Figure 2.7. Probability of transmission rates as a function of average SNR ρ [(a): $SNR_2 = 10dB, SNR_{5.5} = 11.5dB, SNR_{11} = 12dB$; (b): $SNR_2 = 10dB, SNR_{5.5} = 15dB, SNR_{11} = 20dB$; (c): $SNR_2 = 10dB, SNR_{5.5} = 20dB, SNR_{11} = 30dB$; (d): $SNR_2 = 10dB, SNR_{5.5} = 30dB, SNR_{11} = 60dB$]

2.7.3. Example 3: The Continuous Rate Case for Heterogeneous Networks

Based on (2.18), it can be shown that the optimal threshold for the heterogeneous case, x^*, satisfies the following equation:

$$x^* = \frac{1}{\delta} \sum_{m=1}^{M} p_{s,m} \exp\left(\frac{1}{\rho_m}\right) E_1\left(\frac{\exp(x^*)}{\rho_m}\right). \tag{2.63}$$

Note that the average throughput without using optimal stopping rule is given by

$$x^L = \frac{\sum_{m=1}^{M} p_{s,m} \exp(1/\rho_m) E_1\left(1/\rho_m\right)}{\delta + p_s}. \tag{2.64}$$

In the following example, we consider a heterogeneous network model with 5 users, each with different transmission probabilities and channel statistics. The performance of the iterative algorithm (2.20) is examined in Table 2.4. Clearly, the iterative algorithm (2.20) exhibits fast convergence rate.

As is clear in (2.63), the optimal threshold x^* (namely the maximum throughput) depends on all SNR parameters $\{\rho_m, \forall m\}$ across links, and is monotonically increasing in

Table 2.4. Convergence behavior of the iterative algorithm (2.20)

ρ (dB)	x_0	x_1	x_2	x_3	x^*
[0 10 10 8.5 6]	0.684	1.259	1.382	1.385	1.385
[10 10 10 8.5 6]	0.026	1.620	1.877	1.892	1.892
[20 10 10 8.5 6]	0.777	2.695	3.054	3.073	3.073

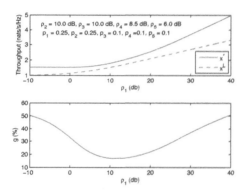

Figure 2.8. Throughput gain $g(\rho)$ as a function of average SNR ρ_1

each ρ_m. However, different from the homogeneous case, the gain g is no longer monoton-ically decreasing in each individual SNR. To get a more concrete sense, we plot in Fig. 2.8 the relationship between g and ρ_1, with other SNR parameters fixed. As illustrated in the figure, g decreases as ρ_1 increases from -10dB to 10dB. This is because when ρ_1 is small, the optimal throughput x^* is determined mainly by other SNR parameters and remains almost constant, whereas the throughput without using optimal stopping strategy (x^L) always increases. Furthermore, g increases when ρ_1 exceeds 10dB. Our intuition is that in this SNR regime user 1 becomes the dominating user in the system, and therefore x^* increases much faster than x^L.

2.8. Conclusions

In this chapter, we considered ad-hoc communications based on random access, and studied distributed opportunistic scheduling to resolve collisions therein while exploiting channel variation. In ad-hoc communications, distributed opportunistic scheduling boils down to a process of joint channel probing and distributed scheduling, and we investigated distributed opportunistic scheduling from a network-centric point of view, where links co-operate to maximize the overall network throughput. Specifically, we treated the channel probing and scheduling as a maximal-rate-of-return problem, and characterized the optimal strategy that yields the maximum overall throughput, for both homogenous networks and heterogeneous networks. We showed that the optimal strategy is a pure threshold policy, where the threshold is the solution to a fixed point equation. Furthermore, we devised an iterative algorithm to compute the optimal threshold. Simulation results indicate that the distributed opportunistic scheduling achieves significant gain provided that the wireless channel exhibits sufficient fluctuation.

CHAPTER 3

DISTRIBUTED OPPORTUNISTIC SCHEDULING UNDER NOISY CHANNEL

ESTIMATION

3.1. Introduction

In the previous chapter, we have developed distributed opportunistic scheduling (DOS)

to exploit multiuser diversity and time diversity for network throughput optimization. A

key assumption in the previous study is that the channel estimation is perfect so that the

channel state information (CSI) is perfectly known at the receiver/transmitter. However,

in practical scenarios, channel conditions are often estimated using noisy observations.

Therefore, it is also of great interest to study DOS under noisy channel estimation. In

such cases, it turns out that the estimated SNR is always larger than the "actual SNR".

Thus, if the data were transmitted using the estimated SNR, there would always be an

outage. To reduce the outage probability, the transmitted rate has to back off from

the estimated rate. Therefore, different from the perfect estimation case, the optimal

scheduling policy hinges on the backoff rate.

In this chapter, we show that the optimal scheduling policy has a threshold structure,

but the threshold turns out to be a function of the variance of the estimation error, and

furthermore it is a functional of the backoff rate. Since the optimal backoff rate function

(in terms of maximizing the network throughput) is intractable, we propose a suboptimal

scheme with a backoff ratio σ. We show that the corresponding optimal backoff ratio

and the corresponding threshold can be obtained via an iterative algorithm. Simulation

results are provided to show that the distributed opportunistic scheduling achieves sig-

[1] ⓒ [2009] IEEE. Reprinted, with permission, from Dong Zheng, Man-on Pun, Weiyan Ge, Junshan Zhang and Vicent Poor, "Distributed Opportunistic Scheduling For Ad Hoc Communications With Imperfect Channel Information", IEEE Trans. on Wireless Comm., Vol. 7, Issue 12, Part 2, December 2008 Pages(s): 5450-5460.

nificant throughput gain in the presence of noisy channel estimation, especially in the low SNR region. In addition, we observe that the performance loss of the distributed opportunistic scheduling, due to the imperfect channel estimation, is less than that of the schemes without using channel-aware scheduling, indicating that the proposed distributed opportunistic scheduling is more robust to noisy channel estimation. Clearly, longer training would increase the accuracy of the channel estimation, but it also decreases the time for data transmission. We use simulation to illustrate this tradeoff.

Perhaps [92] is the most related work to our study here, where in [92] the effect of the channel estimation error on centralized opportunistic scheduling has been studied. The major difference between this study and the study in [92] is that we consider distributed opportunistic scheduling for ad hoc communications under noisy channel estimation where the noisy channel estimation is available only after a successful channel contention, while [92] considers centralized scheduling assuming the noisy channel estimations of all the links are available at the base station before the scheduling.

The rest of the chapter is organized as follows. We present the system model in Section 3.2. We then study DOS under noisy channel estimation in Section 3.3. The numerical results are presented in Section 3.4. Finally, we draw our conclusions and discuss future work in Section 3.5.

3.2. System Model

We consider the same network model as described in Section 2.3 in Chapter 2.

Next, we develop the signal model for the channel estimation. Let $s(n)$ denote the successful link at the n-th successful channel contention. The corresponding received signal is given by:

$$Y_{s(n)}(n) = \sqrt{\rho}h_{s(n)}(n)X_{s(n)}(n) + \mu_{s(n)}(n), \tag{3.1}$$

where ρ is the normalized receiver SNR, $h_{s(n)}(n)$ is the channel coefficient for link $s(n)$, $X_{s(n)}(n)$ is the transmitted signal with $E||X_{s(n)}(n)||^2 = 1$ and $\mu_{s(n)}(n)$ is the additive white noise with i.i.d. $\mathcal{CN}(0,1)$.

To simplify the exposition, we consider a homogeneous network where all links have the same channel statistics, and are subject to Rayleigh fading, i.e., $h_{s(n)}(n)$ follows a complex Gaussian distribution $\mathcal{CN}(0,1)$. In what follows, we drop the subscripts to simplify the notation and use h_n to stand for $h_{s(n)}(n)$ where it is clear from the context. Similarly, we use Y_n, X_n and μ_n to denote $Y_{s(n)}(n)$, $X_{s(n)}(n)$ and $\mu_{s(n)}(n)$.

We consider the continuous rate case, assuming that the instantaneous rate is given by the Shannon channel capacity, i.e.,

$$R_n = \log(1 + \rho|h_n|^2) \text{ nats/s/Hz},$$

provided that the channel can be perfectly estimated.

3.3. Distributed Opportunistic Scheduling Under Noisy Channel Estimation

Needless to say, in practical systems, h_n has to be estimated using training signals (e.g. embedded in the RTS packets). Let $\hat{h}_n$ denote the estimation of the channel coefficient, and $\tilde{h}_n$ the estimation error. It follows that

$$h_n = \hat{h}_n + \tilde{h}_n, \tag{3.2}$$

where $\hat{h}_n$ and $\tilde{h}_n$ are zero-mean complex Gaussian random variables. Suppose that the channel is estimated using an MMSE estimator. It follows, by the orthogonality principle, that

$$E[|h_n|^2] = E[|\hat{h}_n|^2] + E[|\tilde{h}_n|^2].$$ (3.3)

Let β denote the variance of the estimation error, and from (3.3), we have that

$$E[|\tilde{h}_n|^2] = \beta,$$

$$E[|\hat{h}_n|^2] = 1 - \beta.$$

Treating the estimation error as noise, the actual SNR at the receiver can be computed by [92]

$$\lambda_n = \frac{\rho|\hat{h}_n|^2}{1 + \rho|\tilde{h}_n|^2}.$$ (3.4)

We note that the numerator of (3.4), $\rho|\hat{h}_n|^2$, is the estimated SNR. Therefore, in contrast to the perfect CSI case where the sequence $\{\rho|h_n|^2, n = 1, 2, \ldots\}$ is used for distributed scheduling, in the noisy channel estimation case, $\{\rho|\hat{h}_n|^2, n = 1, 2, \ldots\}$ serves as the basis for distributed scheduling, i.e.,

$$\{N = n\} \in \mathcal{F}'_n,$$ (3.5)

where $\mathcal{F}'_n$ is the σ-algebra generated by $\{(\rho|\hat{h}_j|^2, K_j), j = 1, 2, \ldots, n\}$.

Following [92], $|\hat{h}_n|^2$ and $|\tilde{h}_n|^2$ can be normalized as

$$\hat{\lambda}_n = \frac{|\hat{h}_n|^2}{1 - \beta}, \quad z_n = \frac{|\tilde{h}_n|^2}{\beta}.$$ (3.6)

Note that both $\hat{\lambda}_n$ and z_n have exponential distribution with unit variance. Furthermore, λ_n in (3.4) can be rewritten as

$$\lambda_n = \frac{\rho_{eff}\hat{\lambda}_n}{1 + \alpha\rho_{eff}z_n},$$ (3.7)

where $\rho_{eff} \triangleq (1 - \beta)\rho$ and $\alpha \triangleq \frac{\beta}{1-\beta}$ denote the "effective channel SNR" and "normalized error variance", respectively. It can be shown the distribution of λ_n given $\hat{\lambda}_n$ is given by [92]

$$f\left(\lambda_n | \hat{\lambda}_n\right) = \exp\left\{ -\frac{1}{\alpha}\left(\frac{\hat{\lambda}_n}{\lambda_n} - \frac{1}{\rho_{eff}}\right)\right\} \frac{\hat{\lambda}_n}{\alpha \lambda_n^2} \mathbf{I}\left(\frac{\hat{\lambda}_n}{\lambda_n} - \frac{1}{\rho_{eff}}\right), \tag{3.8}$$

where $\mathbf{I}(\cdot)$ is the indicator function.

3.3.1. Optimal Distributed Scheduling under Noisy Channel Estimation

In this section, we explore optimal scheduling for the noisy channel estimation case.

It is clear that the actual SNR λ_n is no greater than the estimated SNR $\rho_{eff}\hat{\lambda}_n$. As a result, if the packet is transmitted at the estimated rate $\log(1 + \rho_{eff}\hat{\lambda}_n)$, there will always be a channel outage. Therefore, the transmission rate has to back off from the estimate rate. Equivalently, we can back off the estimated SNR $\rho_{eff}\hat{\lambda}$ to a "nominated" SNR $\lambda_c(\hat{\lambda})$. Accordingly, the instantaneous rate, R_n, is given by

$$R_n = \log\left(1 + \lambda_c(\hat{\lambda}_n)\right) \mathbf{I}\left(\lambda_c(\hat{\lambda}_n) \le \lambda_n\right). \tag{3.9}$$

Along the same line as in the perfect CSI case, for each given back-off rate function $\lambda_c(\cdot)$, maximizing the average throughput boils down to solving the maximal rate of return problem in (2.1), i.e.,

$$N^* \triangleq \arg\max_{N \in Q} \frac{E[R_N T]}{E[T_N]}, \tag{3.10}$$

where R_n is given in (3.9).

Observe that there are at least two major differences between the perfect estimation case and the noisy channel estimation case. First, the stopping rule N is now defined by the σ-field $\mathcal{F}'_n$, instead of $\mathcal{F}_n$. Second, the instantaneous rate, R_n, defined in (3.9), is now a random variable, and is not perfectly known at time n. However, it can be shown that the structure of the optimal scheduling strategy remains the same if the random "reward"

R_n is replaced with its conditional expectation, denoted as $\bar{R}_n$ [39, Page 1.3]. As a result, the scheduling can now be based on $\bar{R}_n$ instead of R_n, where

$$
\begin{aligned}
\bar{R}_n &= E\left[R_n | \mathcal{F}_n'\right] \\
&= E\left[\log\left(1 + \lambda_c(\hat{\lambda}_n)\right) \mathbf{I}\left(\lambda_c(\hat{\lambda}_n) \le \lambda_n\right) | \mathcal{F}_n'\right] \\
&= \log\left(1 + \lambda_c(\hat{\lambda}_n)\right) P\left(\lambda_c(\hat{\lambda}_n) \le \lambda_n | \mathcal{F}_n'\right) \\
&= \log\left(1 + \lambda_c(\hat{\lambda}_n)\right) \int_{\lambda_c(\hat{\lambda}_n)}^{\rho_{eff}\hat{\lambda}_n} \exp\left\{-\frac{1}{\alpha}\left(\frac{\hat{\lambda}_n}{\lambda} - \frac{1}{\rho_{eff}}\right)\right\} \frac{\hat{\lambda}_n}{\alpha\lambda^2} d\lambda \\
&= \log\left(1 + \lambda_c(\hat{\lambda}_n)\right) \left[1 - \exp\left\{-\frac{1}{\alpha}\left(\frac{\hat{\lambda}_n}{\lambda_c(\hat{\lambda}_n)} - \frac{1}{\rho_{eff}}\right)\right\}\right],
\end{aligned}
$$

where we have used the fact that $P\left(\lambda_c(\hat{\lambda}_n) \le \lambda_n | \mathcal{F}_n'\right) = P\left(\lambda_c(\hat{\lambda}_n) \le \lambda_n | \hat{\lambda}_n\right)$ due to the independence of channel estimations.

Based on the above analysis and the results in Chapter 2, we conclude that the optimal scheduling policy under noisy channel estimation is still a pure threshold policy, where the optimal threshold x^* can be computed from (2.5), and it is the unique solution to the following fixed point equation

$$x = \Phi(x, \lambda_c), \tag{3.11}$$

where

$$
\Phi(x, \lambda_c) \triangleq \frac{\int_{\hat{\lambda}'}^{\infty} e^{-\hat{\lambda}} \log\left(1 + \lambda_c\right) \left[1 - \exp\left\{-\frac{1}{\alpha}\left(\frac{\hat{\lambda}}{\lambda_c} - \frac{1}{\rho_{eff}}\right)\right\}\right] d\hat{\lambda}}{\frac{\delta}{p_s} + e^{-\hat{\lambda}'}}, \tag{3.12}
$$

with $\hat{\lambda}'$ defined as

$$
\log\left(1 + \lambda_c(\hat{\lambda}')\right) \left[1 - \exp\left\{-\frac{1}{\alpha}\left(\frac{\hat{\lambda}'}{\lambda_c(\hat{\lambda}')} - \frac{1}{\rho_{eff}}\right)\right\}\right] = x. \tag{3.13}
$$

3.3.2. Optimal Backoff Rate Function

It is clear from that (3.11) that for any given backoff rate function $\lambda_c(\cdot)$, there is a corresponding optimal throughput x^*. Therefore, x^* is a functional of $\lambda_c(\cdot)$, denoted as

$x^*(\lambda_c)$. We are interested to find the function $\lambda_c^*(\cdot)$ that maximizes $x^*(\lambda_c)$, i.e.,

$$\lambda_c^* = \arg\max_{\lambda_c \in A} x^*(\lambda_c), \tag{3.14}$$

where A is the set of the admissible functions (for example, A can be $\{\lambda_c(\hat{\lambda}) : \lambda_c(\hat{\lambda}) \geq 0, \forall\, \hat{\lambda} \geq 0\}$).

Based on the theory of calculus of variations [40], problem (3.14) is a *variational problem*, and the functions $\lambda_c^*(\cdot)$ are called *extremals*. However, different from the canonical calculous of variations problems, in this problem, the functional x^* is not explicitly defined on λ_c. Instead, they are connected through a fixed point equation. Furthermore, the integral range in (3.12) is not fixed, but is a function of λ_c (cf. (3.13)). As a result, it is intractable to characterize λ_c^*.

3.3.3. Suboptimal Multiplicative Backoff Rate Function

In what follows, we propose a suboptimal backoff rate function, which backs off the estimated SNR by a multiplicative ratio σ, i.e., we set

$$\lambda_c(\hat{\lambda}) = \sigma \rho_{eff} \hat{\lambda}, \tag{3.15}$$

and $0 \leq \sigma \leq 1$.

It follows from (3.15), (3.12) and (3.13) that the optimal throughput x^* is the solution to

$$x = \Phi(x,\sigma)$$

$$= \left[1 - \exp\left\{-\frac{1}{\alpha\rho_{eff}}\left(\frac{1}{\sigma}-1\right)\right\}\right] \frac{\int_{\hat{\lambda}'}^{\infty} \exp(-\hat{\lambda}) \log\left(1 + \sigma\rho_{eff}\hat{\lambda}\right) d\hat{\lambda}}{\frac{\delta}{p_s} + \exp(-\hat{\lambda}')}$$

$$= \left[1 - \exp\left\{-\frac{1}{\alpha\rho_{eff}}\left(\frac{1}{\sigma}-1\right)\right\}\right]$$

$$\times \frac{\log(1 + \sigma\rho_{eff}\hat{\lambda}')e^{-\hat{\lambda}'} + \exp\left(\frac{1}{\sigma\rho_{eff}}\right) E_1\left(\hat{\lambda}' + \frac{1}{\sigma\rho_{eff}}\right)}{\frac{\delta}{p_s} + e^{-\hat{\lambda}'}}, \qquad (3.16)$$

where

$$\hat{\lambda}' = \frac{\exp\left(\frac{x}{1-\exp\left\{-\frac{1}{\alpha\rho_{eff}}\left(\frac{1}{\sigma}-1\right)\right\}}\right) - 1}{\sigma\rho_{eff}}. \qquad (3.17)$$

It is not difficult to show that x^* is a continuous and differentiable function of σ, and hence, there exists an optimal backoff ratio σ^* such that

$$\sigma^* = \arg\max_{\sigma} x^*(\sigma). \qquad (3.18)$$

It can also be shown that σ^* cannot be 0 or 1 (since the corresponding throughput is zero). Therefore, the optimal ratio σ^* must satisfy the first order condition $\frac{dx^*(\sigma^*)}{d\sigma} = 0$.

3.3.4. An Iterative Algorithm for Computing σ^* and $x^*(\sigma^*)$

Due to the complicated structure of the fixed point equation (3.16), it is not feasible to characterize σ^* using the first order condition. In what follows, we devise an iterative algorithm instead based on fractional optimization techniques [13].

Specifically, we define the following functions:

$$U(\sigma,x) \triangleq \left[1 - \exp\left\{-\frac{1}{\alpha\rho_{eff}}\left(\frac{1}{\sigma}-1\right)\right\}\right] \times$$
$$\left\{\log(1 + \sigma\rho_{eff}\hat{\lambda}')\exp(-\hat{\lambda}') + \exp\left(\frac{1}{\sigma\rho_{eff}}\right) E_1\left(\hat{\lambda}' + \frac{1}{\sigma\rho_{eff}}\right)\right\} \qquad (3.19)$$

$$V(\sigma,x) \triangleq \frac{\delta}{p_s} + \exp(-\hat{\lambda}'), \qquad (3.20)$$

where $\hat{\lambda}'$ is defined in (3.17).

We propose the following iterative algorithm:

Algorithm 1 Iterative Algorithm for Computing σ^* and $x^*(\sigma^*)$

States:
  x_k, σ_k
Init:
  x_0
Procedure:
  while $|x_k - x_{k-1}| > \epsilon$ **do**
    $\sigma_{k-1} = \arg\max\limits_{0 \le \sigma \le 1} \{U(\sigma, x_{k-1}) - x_{k-1}V(\sigma, x_{k-1})\}$
    $x_k = \frac{U(\sigma_{k-1}, x_{k-1})}{V(\sigma_{k-1}, x_{k-1})}$
  end while

We have the following proposition regarding the convergence of Algorithm 1.

Proposition 3.3.1. *The iterates* $\{(\sigma_k, x_k), k = 1, 2, \ldots\}$ *generated by Algorithm 1 converge to* $(\sigma^*, x^*(\sigma^*))$ *for any positive value* x_0.

Proof: To maximize the throughput $U(\sigma, x)/V(\sigma, x)$, we construct the function

$$W(\sigma, x, y) = U(\sigma, x) - yV(\sigma, x), \tag{3.21}$$

where y is a real number. For a given y, it is not difficult to see that there exist a unique pair $\sigma(y)$ and $x(y)$ that maximize $W(\sigma, x, y)$ (note that $\sigma(y)$ and $x(y)$ are well defined by the uniqueness). Let $W(y) \triangleq W(\sigma(y), x(y), y)$. It then can be shown that y^* such that $W(y^*) = 0$ is the optimal throughput [13], i.e., $y^* = x^*(\sigma^*) = x(y^*)$ and $\sigma^* = \sigma(y^*)$.

To obtain y^*, we first observe that $W(y)$ is *decreasing* and *convex* in y. We can also show that for any given y_0, $s = W(y_0) - V(\sigma(y_0), x(y_0))(y - y_0)$ is a supporting hyperplane for $W(y)$ at y_0. It follows that from Newton's method, we have the following iterative algorithm for approximating y^*:

$$
\begin{aligned}
y_{k+1} &= y_k - \frac{W(y_k)}{W'(y_k)} \\
&= y_k + \frac{W(y_k)}{V(\sigma(y_k), x(y_k))} \\
&= \frac{U(\sigma(y_k), x(y_k))}{V(\sigma(y_k), x(y_k))}.
\end{aligned}
\tag{3.22}
$$

By the definition of $U(\cdot, \cdot)$, $V(\cdot, \cdot)$ and $W(\sigma, x, y)$, it can be shown that

$$
x(y) = y.
\tag{3.23}
$$

Therefore, we can replace y_k by x_k, and the iterative algorithm (3.22) becomes

$$
x_{k+1} = \frac{U(\sigma(x_k), x_k)}{V(\sigma(x_k), x_k)}.
\tag{3.24}
$$

where

$$
\sigma(x_k) = \arg\max_{\sigma}\{U(\sigma, x_k) - x_k V(\sigma, x_k)\}.
\tag{3.25}
$$

According to Newton's method, given any initial value x_0, the above iterative algorithms (3.24) and (3.25) are known to converge quadratically to the optimal network throughput $x^*(\sigma^*)$ and the optimal backoff ratio σ^* [13]. ∎

3.4. Numerical Results

In this section, we provide numerical examples to illustrate the above results. Unless otherwise specified, we assume that τ, T, p, and m are chosen such that $\delta = 0.1, p_s = \exp(-1)$.

Figure 3.1 depicts $\Phi(x, \sigma)$ as a function of the backoff ratio σ. It can be seen that the average throughput is zero at both $\sigma = 0$ and $\sigma = 1$, and is maximized somewhere in between.

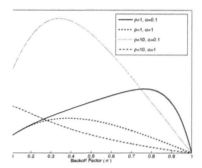

Figure 3.1. $\Phi(\sigma)$ vs. σ, $x = 0.1$.

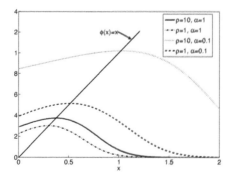

Figure 3.2. $\Phi(x, \sigma^*)$ vs. x

Table 3.1. Convergence behavior of the iterative algorithm, $\alpha = 1$.

ρ	x_0	x_1	x_2	x_3	x^*	σ^*
0.5	0.5	0.177	0.246	0.254	0.254	0.407
1	0.5	0.254	0.299	0.301	0.301	0.285
2	0.5	0.306	0.335	0.336	0.336	0.182
5	0.5	0.344	0.363	0.364	0.364	0.090
10	0.5	0.358	0.374	0.374	0.374	0.049

Figure 3.2 depicts $\Phi(x, \sigma^*)$ as a function of x. Note that the optimal throughput x^* is the solution to the fixed point equation $x = \Phi(x, \sigma^*)$. It can be observed that x^* is an increasing function of ρ for a given α, and is a decreasing function of α for a fixed ρ. It can also be seen that the estimation accuracy plays an important role in the throughput performance: when α decreases from 1 to 0.1, the performance improves almost 70% for $\rho = 10$.

In Table 3.1, we examine the convergence of the iterative algorithm I with $\alpha = 1$. As expected, $x(n)$ approaches to x^* usually within a few iterations.

Table 3.2 compares the convergence behavior of the iterative algorithm with different error variance α, where the SNR is fixed as $\rho = 1$. When the error variance is large, the iterative algorithm needs more iterations to converge. Moreover, the backoff ratio σ would decrease as α increases. This can be further observed in Fig. 3.3. It indicates that when the estimation error is large, the transmitter would backoff more to avoid channel outage.

Table 3.3 illustrates the throughput gain

$$g = \frac{x^* - x^L}{x^L} \tag{3.26}$$

as a function of ρ, where $x^L = \Phi(0, \sigma^*)$ is the average throughput obtained by the schemes

Table 3.2. Convergence behavior of the iterative algorithm, $\rho = 1$.

α	x_0	x_1	x_2	x_3	x_4	x_5	x^*	σ^*
0	0.5	0.604	0.610				0.610	1.00
0.1	0.5	0.514	0.514				0.514	0.753
1	0.5	0.254	0.299	0.301			0.301	0.285
2	0.5	0.109	0.201	0.217	0.218		0.218	0.155
5	0.5	0.004	0.091	0.120	0.122	0.123	0.123	0.054

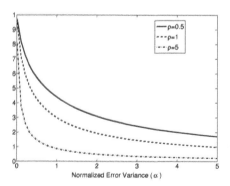

Figure 3.3. Backoff factor σ as a function of normalized error variance α

Table 3.3. Throughput gain of DOS, $\alpha = 1$.

ρ	0.5	1	2	5	10	100
x^*	0.254	0.301	0.336	0.364	0.374	0.385
x^L	0.185	0.224	0.254	0.278	0.288	0.298
$g(\rho)$	37.3%	34.3%	32.3%	30.9%	29.8%	29.2%

Table 3.4. Throughput gain of DOS, $\rho = 0.5$.

α	0	0.01	0.1	1	2	5
x^*	0.384	0.378	0.352	0.254	0.197	0.118
x^L	0.284	0.279	0.259	0.186	0.143	0.085
$g(\alpha)$	35.2%	35.5%	35.9%	36.6%	37.8%	38.8%

without using optimal scheduling. It can be seen that the throughput gain is more significant in the low SNR region, and is a decreasing function of ρ.

In Table 3.4, we illustrate the throughput gain as a function of α. Note that $\alpha = 1/(1 - \beta) - 1$ is an increasing function of β. As expected, when the normalized noise variance α increases, the optimal throughput x^* decreases, as well as the x^L. However, it is interesting to observe that the throughput gain increases instead. The rationale behind is that the performance of the schemes without using optimal scheduling "suffers" more than that of the distributed opportunistic scheduling in the presence of noisy channel estimation.

We also examine the performance of the distributed opportunistic scheduling against the training time τ. According to linear estimation theory [48], the error variance β and the training time τ has the following relationship:

$$\beta = \frac{1}{\rho\tau + 1}. \tag{3.27}$$

Using (3.27) in the simulation, we plot throughput performance of the distributed oppor-

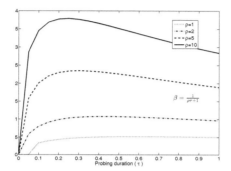

Figure 3.4. Throughput vs. training time τ

tunistic scheduling against the training time τ in Fig. 3.4. It is clear that there exists an optimal training time which balances the tradeoff between better estimation accuracy and loss of transmission time. It can also be observed that when the average SNR ρ increases, the optimal training time decreases.

3.5. Conclusion

In this chapter, we generalized the study in Chapter 2, and considered DOS under noisy channel estimation. For such cases, we proposed that the transmitted rate backs off on the estimated rate so as to reduce the channel outage probability. We showed that the optimal scheduling policy is still threshold-based, but the threshold turns out to be a function of the variance of the estimation error, and is a functional of the backoff rate. Since the optimal backoff is intractable, we proposed a suboptimal scheme that backs off on the estimated SNR (and hence the rate) through a backoff ratio σ. The corresponding optimal backoff ratio and the optimal threshold can be obtained using an iterative algorithm.

Simulation results indicate that the distributed opportunistic scheduling still achieves significant throughput gain in the presence of noisy channel estimation, especially in the low SNR region. In addition, we observed that the performance loss of the distributed opportunistic scheduling due to the imperfect channel estimation is less than that of the schemes without using optimal scheduling, indicating that the devised distributed opportunistic scheduling is more robust to the noisy channel estimation. Finally, we used simulation to find the optimal training time.

CHAPTER 4

DISTRIBUTED OPPORTUNISTIC SCHEDULING FOR EXPLOITING

MULTI-RECEIVER DIVERSITY AND MULTIUSER DIVERSITY

4.1. Introduction

4.1.1. Motivation

In Chapter 2, we have shown that significant channel diversity gain can be obtained using distributed opportunistic scheduling, assuming that each transmitter only has one receiver. However, in many wireless applications, it is very likely that a node may have access to multiple channels [80, 41, 30] or multiple intended receivers [47, 28]. For example, the existing IEEE 802.11a/b standards have already specified multiple frequencies for data communications, and the cutting-edge OFDM and MIMO technologies can serve multiple receivers in future wireless systems. To fully exploit the multi-receiver/multiuser diversities gives rise to significant challenges for upper-layer protocol design. Particularly, it is of great interest to leverage such degrees of freedom for MAC layer design. One unique challenge in exploiting multi-receiver/multiuser diversities is that in random-access based ad-hoc networks, probings (contentions) are needed to reserve the channel and also to track and discover channel conditions, so that data transmissions could be carried out over favorable channel conditions. Recent studies [80, 41, 30] have explored channel probing for a single link system with multiple channels, and the schemes therein are designed for point-to-point communications. In contrast, in this study, we consider ad-hoc networks with many links, and the focus here is to develop optimal channel-aware distributed scheduling for network throughput optimization by leveraging multi-receiver/multiuser diversities and time diversity in a distributed manner [105].

4.1.2. Network model

We consider a single-hop ad-hoc network with M transmitter nodes, each with multiple intended receivers. We assume that each transmitter node m contends for the channel using random access with probability p_m, $m = 1, \ldots, M$. A collision model is assumed for channel contention, where a channel contention of a node is said to be successful if no other nodes transmit at the same time. Accordingly, the overall successful contention probability is given by $\sum_{m=1}^{M} (p_m \prod_{i \neq m} (1 - p_i))$, denoted as p_s; and the number of slots (denoted as K) needed to accomplish a successful channel contention is a Geometric random variable, i.e., $K \sim Geometric(p_s)$. Let τ denote the duration of mini-slot for channel contention, and T the duration of data transmission.

Different from the model where each transmitter is associated with one single receiver only, the probing in the multi-receivers case takes place in two phases (see Fig. 4.4 for example): 1) In phase I, each transmitter contends for the channel using random access to reserve the channel (e.g., by sending RTS), and the probing in this phase to accomplish a successful channel contention takes a *random duration of $K\tau$*; and 2) In phase II, subsequent probings are carried out to estimate the channel conditions to different intended receivers of this successful transmitter, according to specific probing mechanisms, and for each receiver the probing for channel condition incurs a *constant duration of τ*. In particular, we shall study four probing mechanisms, namely, 1) the random selection (RS) mechanism, 2) the exhaustive sequential probing with recall (ESPWR) mechanism, 3) the sequential probing without recall (SPWOR) mechanism, and 4) the sequential probing with recall (SPWR) mechanism.

Suppose that after channel probing, the link condition of the probed receiver can be obtained accurately. Due to channel fading, the link condition corresponding to a probed receiver can be either good or bad (we shall make this precise). Clearly, the chance of seeing better channel conditions increases if further probing is performed. However, each probing incurs a certain amount of time that could be used for data transmission. Therefore, there exist fundamental tradeoffs between the throughput gain from better channel conditions and the probing cost. In this chapter, we take a systematic approach to characterize this tradeoff by appealing to optimal stopping theory [35, 39], and explore channel-aware distributed scheduling to exploit multi-receiver/multiuser diversities for ad-hoc communications. For all four probing mechanisms, we characterize the corresponding optimal scheduling policies; and show that the optimal scheduling boils down to joint execution of channel probing using an optimal stopping rule and then data transmission.

4.1.3. Summary of Main Results

We study channel aware distributed scheduling for both unicast traffic and multicast traffic.

1) For the unicast case, we show that the optimal scheduling policies for all four probing mechanisms exhibit threshold structures, and that the stopping decisions are based on the thresholds and local channel conditions. Particularly, we show that for the RS and ESPWR probing mechanisms, the corresponding optimal scheduling strategies are single threshold policies, where the maximum throughput is an optimal threshold. In contrast, the optimal scheduling policies for both SPWOR and SPWR are multi-stage threshold policies, where the optimal thresholds are functions of the number of probed receivers. Furthermore, we show that the optimal thresholds and the maximum throughput can be

obtained *off-line* by solving fixed point equations. Therefore, the optimal scheduling policies are amenable to easy distributed implementation. Interestingly, we observe that the optimal thresholds for SPWOR monotonically decrease, whereas the optimal thresholds for SPWR first increase and then decrease.

2) For multicast traffic, we show that the probing process can be treated the same as the ESPWR mechanism. As a result, the optimal scheduling developed for ESPWR is applicable to the multicast case under consideration. Needless to say, the optimal thresholds depend on the specific rewards of interest, and we study two different cases: 1) the reward is the number of ready users; and 2) the reward is the sum rate.

3) We develop iterative algorithms to obtain the optimal thresholds by appealing to the technique of fractional maximization [13]. We establish the convergence of the iterative algorithms, and show that the convergence rate is quadratic. Particularly, we derive iterative algorithms to compute the optimal thresholds for the four probing mechanisms for the continuous rate case.

The rest of the chapter is organized as follows. In Section 4.2, we study channel aware distributed scheduling for two important applications, namely, the unicast traffic and the multicast traffic. In particular, Section 4.2.1, the main focus of this study, presents the problem formulation for the unicast application, and investigates the optimal scheduling policies for the four probing mechanisms. Section 4.2.3 presents the study for the multicast application. In Section 4.3, we develop iterative algorithms for computing the optimal thresholds and the maximum throughput. The numerical examples in Section 4.4 corroborate the theoretic findings. Finally, Section 4.5 concludes the chapter.

4.2. Channel Aware Distributed Scheduling for Exploiting Multi-Receiver Diversity and Multiuser Diversity

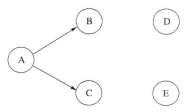

Figure 4.1. A sketch of a transmitter with multiple receivers.

In this section, we generalize the above study in Chapter 2 to the network model where each transmitter node has multiple receivers (see Fig. 4.1), and the objective is to maximize the overall network throughput. For ease of exposition, we first consider a homogeneous network where all transmitters have the same number (say L) of receivers[1], and the channel condition follows the same distribution $F(r)$. We will generalize the study to heterogenous networks in Section 4.2.2. Without loss of optimality, we assume that the probing order for each transmitter is based on the numbering of the receivers, numbered 0 to $L - 1$.

Let $t(n)$ denote the transmitter accomplishing a successful channel contention in the n-th round, and $R_{n,t(n),j}$ be the corresponding rate after receiver j is probed, $j = 0, 1, \ldots, L - 1$. In wireless communications, $R_{n,t(n),j}$ depends on the time varying channel condition. Following the standard assumption on block fading in wireless communications [44], we assume that the rate $R_{n,t(n),j}$ remains constant for a duration of $(L - 1)\tau + T$, which is no greater than the channel coherence time. Without loss of generality, we impose the

---

[1]It is known that most multi-receiver gain occurs for $L = 2$ and $L = 3$.

following assumption on the transmission rates:

A1) $\{R_{n,t(n),j}\}$ are i.i.d., and $E[R^2_{n,t(n),j}] < \infty, \forall\, n, j$.

A key observation is that the probing costs (in terms of the probing time) for acquiring $R_{n,t(n),0}$ and $R_{n,t(n),j}, j = 1, 2, \ldots, L - 1$ are different: It takes a random duration of contention period of $K\tau$ to obtain $R_{n,t(n),0}$, whereas it takes only a constant time τ to obtain $R_{n,t(n),j}, j = 1, 2, \ldots, L - 1$. For convenience, we have assumed that a complete handshake (e.g., RTS/CTS) is used to obtain $R_{n,t(n),j}, j = 1, 2, \ldots, L - 1$. (This can be improved further, e.g., by combining multiple RTS packets into a single multicast RTS packet, and letting the receivers send back CTS packets sequentially.)

4.2.1. Channel Aware Distributed Scheduling for Unicast Traffic

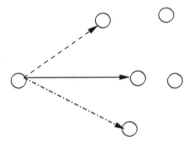

Figure 4.2. A sketch of unicast transmissions.

The main focus of this study is on channel aware distributed scheduling for unicast traffic (see Fig. 4.2). That is, the transmitter transmits to one receiver each time. Clearly, different probing mechanisms lead to different transmission rates and probing costs. In the following, we characterize the optimal scheduling for four probing mechanisms.

4.2.1.1. Mechanism I: Random Selection (RS)

In the random selection (RS) mechanism, the successful transmitter randomly picks one of the receivers to probe, and only the probed receiver sends back the channel condition to the transmitter after the successful channel contention. Accordingly, the optimal scheduling policy is the same as that for the single-receiver case, indicating that the optimal stopping rule, N^* in (2.4) is applicable here and that the optimal throughput x^*_{RS} can be found by solving (2.5).

Since the RS mechanism does not utilize the multi-receiver diversity, it has no advantage over the single-receiver case. The RS mechanism is used as a benchmark for performance comparison with other probing strategies.

4.2.1.2. Mechanism II: Exhaustive Sequential Probing With Recall (ESPWR)

In the mechanism using Exhaustive Sequential Probing With Recall (ESPWR), after a successful channel contention, the corresponding transmitter probes all its receivers sequentially, and the receivers feed back their channel information accordingly. The transmitter then picks the receiver with the best channel condition for possible data transmission. As illustrated in Fig. 4.3, the ESPWR mechanism can be implemented via RTS/CTS handshaking. Suppose that node A wants to probe the channels to the receiver nodes B and C. Node A first sends RTS to node B, and node B estimates the channel condition which is piggybacked on CTS. Upon receiving CTS from node B, node A then sends another RTS to node C which responds with CTS along with its channel information. Finally, node A selects, between node B and node C, the one with better channel condition for data transmissions, provided that the channel condition is "good".

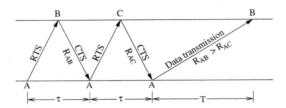

Figure 4.3. ESPWR channel probing using RTS/CTS handshakes.

It follows that the transmission rate and the overall duration for probing plus data transmission are given by

$$R_{n,L} \triangleq \max_{j \in \{0,1,\ldots,L-1\}} R_{n,t(n),j},\qquad(4.1)$$

$$T_n = \sum_{i=1}^{n} [K_i \tau + (L-1)\tau] + T.\qquad(4.2)$$

Proposition 4.2.1. *a) Suppose that exhaustive sequential probing with recall (ESPWR) is used for channel probing. Then the optimal stopping rule for distributed scheduling is given by*

$$N^*_{ESPWR} = \min\{n \geq 1 : R_{n,L} \geq x^*_{ESPWR}\};\qquad(4.3)$$

b) The maximum network throughput x^*_{ESPWR} is an optimal threshold and is the unique solution to

$$E(R'_{n,L} - x)^+ = \frac{x[1 + p_s(L-1)]\delta}{p_s}.\qquad(4.4)$$

Remarks. We caution that the ESPWR mechanism may not always yield higher throughput than the RS mechanism, because the multi-receiver diversity gain would be offset by the probing cost as L increases. In the extreme case, $x^*_{ESPWR} \to 0$ as $L \to \infty$, whereas x_{RS}^* is positive. As a result, there exists an optimal L^*, such that the throughput gain of ESPWR over RS is maximum.

4.2.1.3. Mechanism III: Sequential Probing Without Recall (SPWOR)

In the mechanism using sequential probing without recall (SPWOR), after a successful contention, the transmitter probes its receivers sequentially, and stops the probing process once it probes a "good" channel, followed by data transmission. As in [80], we assume that in this probing mechanism the transmitter cannot recall the previous probed receivers in the sense that the transmitter can schedule data transmission to the most recently probed receiver only.

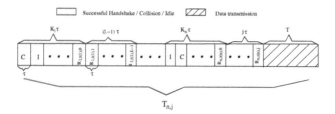

Figure 4.4. A sample realization of the SPWOR mechanism.

As illustrated in Fig. 4.4, suppose that receiver j is selected for data transmission after n rounds of contention. The transmission rate is denoted as $R_{n,t(n),j}$, and the total elapsed time $T_{n,j}$ is given by

$$T_{n,j} = \sum_{i=1}^{n-1} [K_i \tau + (L-1)\tau] + K_n \tau + j\tau + T. \tag{4.5}$$

Clearly, the total number of probed receivers, denoted as $\mathcal{K}$, is

$$\mathcal{K} = (n-1)L + (j+1). \tag{4.6}$$

On the other hand, if the total number of probed receivers $\mathcal{K}$ is known, then n and j would be

$$n = \lceil \frac{\mathcal{K}}{L} \rceil, \qquad j = \mathrm{mod}(\mathcal{K} - 1, L). \tag{4.7}$$

Let $\mathcal{K}(N)$ denote the total number of probed receivers up to a stopping time N. Then, the average throughput, x, is given by

$$x = \frac{E\left[R_{\lceil \frac{\mathcal{K}(N)}{L} \rceil, t(\lceil \frac{\mathcal{K}(N)}{L} \rceil), \mathrm{mod}(\mathcal{K}(N)-1,L)} \times T \right]}{E\left[T_{\lceil \frac{\mathcal{K}(N)}{L} \rceil, \mathrm{mod}(\mathcal{K}(N)-1,L)} \right]}. \tag{4.8}$$

Therefore, the next key step is to characterize the optimal stopping rule N^*, such that x in (4.8) is maximized. i.e.,

$$N^* \triangleq \arg\max_{N \in Q} \frac{E\left[R_{\lceil \frac{\mathcal{K}(N)}{L} \rceil, t(\lceil \frac{\mathcal{K}(N)}{L} \rceil), \mathrm{mod}(\mathcal{K}(N)-1,L)} \times T \right]}{E\left[T_{\lceil \frac{\mathcal{K}(N)}{L} \rceil, \mathrm{mod}(\mathcal{K}(N)-1,L)} \right]}, \tag{4.9}$$

where

$$Q \triangleq \{N : \mathcal{K}(N) \geq 1, E\left[T_{\lceil \frac{\mathcal{K}(N)}{L} \rceil, \mathrm{mod}(\mathcal{K}(N)-1,L)} \right] < \infty\}. \tag{4.10}$$

For convenience, with a slight abuse of notation, we use $\mathcal{K}$ instead of $\mathcal{K}(N)$ in the following.

In general, the optimal stopping rule N^* for the SPWOR mechanism depends on the round of channel contention n and the receiver j, and the corresponding optimal structure is more involved than the single receiver case. However, we shall see that the optimal structure is time invariant in n, and takes a simple threshold form. We have the following result.

Proposition 4.2.2. *a) Suppose that sequential probing without recall (SPWOR) is used for channel probing. Then the optimal stopping rule for distributed scheduling is given as* *follows:*

$$N^*_{SPWOR} = \min\{\kappa \geq 1 : R_{n,t(n),j} \geq \theta^*_j, \ where \ n = \lceil \frac{\kappa}{L} \rceil, j = mod(\kappa - 1, L)\}, \quad (4.11)$$

and the thresholds $\{\theta^*_j\}$ are determined by

$$\theta^*_j = x^*_{SPWOR} + v^*_{j+1}, \forall \ j = 0, 1, \ldots, L - 1; \quad (4.12)$$

b) The maximum network throughput x^*_{SPWOR} is the unique solution to the following fixed point equation:

$$E[\max(R - x, V^*_1(x))] - \frac{x\delta}{p_s} = 0, \quad (4.13)$$

where R is a random variable with distribution $F(r)$, $v^*_j \triangleq V^*_j(x^*_{SPWOR}), \forall \ j = 1, 2, \ldots, L,$ and $\{V^*_j(x)\}$ are defined (in a backward order) as follows:

$$V^*_L(x) \ \triangleq \ 0, \quad (4.14)$$

$$V^*_j(x) \ \triangleq \ E[\max(R - x, V^*_{j+1}(x))] - x\delta, \forall \ j = L - 1, L - 2, \ldots, 1. \quad (4.15)$$

The proof is relegated to Appendix A.1.

Remarks. Proposition 4.2.2 reveals that the optimal scheduling policy corresponding to SPWOR probing exhibits a multi-stage threshold structure. Furthermore, observe that the optimal thresholds given by (4.12) only depend on the number of receivers that the transmitter has probed, indicating that the optimal stopping rule in (4.11) is amenable to easy distributed implementation.

As expected, the optimal thresholds at earlier-probed receivers are larger than that at later-probed receivers, i.e., $\theta^*_i \geq \theta^*_j, \forall \ i \leq j$. Intuitively speaking, at receiver i, more receivers (i.e., $L - i - 1$ remaining receivers) are available for further probing (and can be possibly utilized), compared to at receiver j. The following corollary formalizes this idea.

Corollary 4.2.1. *The optimal thresholds* $\{\theta_j^*, \forall\ j = 0, 1, \ldots, L-1\}$ *defined in (4.12)*

monotonically decrease, i.e.,

$$\theta_0^* \geq \theta_1^* \geq \cdots \geq \theta_{L-1}^*. \tag{4.16}$$

The proof is relegated to Appendix A.2.

4.2.1.4. Mechanism IV: Sequential Probing With Recall (SPWR)

As noted above, the SPWOR mechanism assumes that transmitters cannot recall the previous probed receivers that might have better channel condition than the most recent one. In contrast, in SPWR probing, we assume that each transmitter can schedule data transmission to any of its probed receivers and therefore it would pick from the probed receivers the one with the highest rate. That is, if the current transmitter is $t(n)$, and the current probed receiver is j, then transmitter $t(n)$ can transmit to one of the receivers in $\{0, 1, \ldots j\}$ that has the best condition. We call this mechanism sequential probing with recall (SPWR).

Define $R_{n,j} \triangleq \max(R_{n,t(n),1}, R_{n,t(n),2}, \ldots, R_{n,t(n),j})$. Similar to the case using SPWOR probing, the objective here is to find the optimal stopping rule N^* with

$$N^* \triangleq \underset{N \in Q}{\arg\max} \frac{E\left[R_{\lceil \frac{\kappa}{L} \rceil, \mathrm{mod}(\kappa-1, L)} \times T\right]}{E\left[T_{\lceil \frac{\kappa}{L} \rceil, \mathrm{mod}(\kappa-1, L)}\right]}. \tag{4.17}$$

Proposition 4.2.3. *a) Suppose that sequential probing with recall mechanism (SPWR) is used for channel probing. Then the optimal stopping rule for distributed scheduling is given as follows:*

$$N_{SPWR}^* = \min\{\kappa \geq 1 : R_{n,j} \geq \theta_j^*, \ \textit{where } n = \lceil \tfrac{\kappa}{L} \rceil, j = mod(\kappa - 1, L)\}, \tag{4.18}$$

and the thresholds $\{\theta_j^*\}$ are determined by

$$\theta_{L-1}^* = x_{SPWR}^*, \tag{4.19}$$

$$\theta_j^* = \min\{z : \psi_j^*(z) \leq 0\}, \forall\, j = L-2, L-3, \ldots, 0. \tag{4.20}$$

and $\{\psi_j^*(z), \forall\, j = L-2, L-3, \ldots, 0\}$ are defined as

$$\psi_{L-2}^*(z) \triangleq E\left[(\max\{R-z, x_{SPWR}^* - z\})^+\right] - x_{SPWR}^*\delta, \tag{4.21}$$

$$\psi_j^*(z) \triangleq E\left[\left(\psi_{j+1}^*\left(\max\{z, R\}\right)\right)^+ + (R-z)^+\right] - x_{SPWR}^*\delta, \forall\, j = L-3, L-4, \ldots, 0. \tag{4.22}$$

where R is a random variable with distribution $F(r)$;

  b) The maximum throughput x_{SPWR}^* is the unique solution to the following fixed point equation:

$$E\left[\max\{R - x, U_1^*(R)\}\right] - \frac{x\delta}{p_s} = 0 \tag{4.23}$$

where $U_1^*(z)$ is iteratively defined as follows:

$$U_{L-1}^*(z) \triangleq E\left[\max\{z, R_{L-1}, x\}\right] - x\delta - x, \tag{4.24}$$

$$U_j^*(z) \triangleq E\left[\max\{z, R_j, U_{j+1}^*(\max\{z, R_j\}) + x\}\right] - x\delta - x, \forall\, j = L-2, L-1, \ldots, 1. \tag{4.25}$$

where $\{R_j, j = 1, 2, \ldots, L-1\}$ are i.i.d. random variables with distribution $F(r)$.

  The proof is relegated to Appendix A.3.

  Remarks. 1) Again, we observe that the optimal thresholds determined by (4.19) and (4.20) are functions of the number of probed receivers only. As a result, the transmitter with reserved channel can decide to transmit or not simply based on the current link condition and the threshold corresponding to the number of probed receivers. 2) It can be shown that SPWR achieves the best throughput performance among the four probing mechanisms.

We have the following result regarding the relationship of the optimal thresholds.

Corollary 4.2.2. *The optimal thresholds $\{\theta_j^*, \forall \, j = 0, 1, \ldots, L-1\}$ determined by (4.19)*

and (4.20) satisfy the following relationship:

$$\theta_{L-1}^* \leq \theta_0^* \leq \theta_1^* \leq \cdots \leq \theta_{L-2}^*. \tag{4.26}$$

The proof is relegated to Appendix A.4.

Remarks. Corollary 4.2.2 reveals that the optimal thresholds monotonically increase

from receiver 0 to receiver $L-2$, and then decrease; and the optimal threshold for last

receiver $L-1$ is the lowest among all the thresholds. This is in sharp contrast to the fact

that the optimal thresholds in SPWOR probing monotonically decrease. Our intuition

is as follows: 1) The monotonic increasing of the initial $L-1$ thresholds in SPWR

is due to the fact that SPWR can recall the previous probed receivers. Therefore, the

transmission rate $R_{n,j}$ is a increasing function of j as probing continues, and consequently,

the corresponding thresholds increases. 2) Note that θ_{L-1}^* is the threshold at which the

transmitters decide to re-contend or not. Since channel contention (the channel probing

in Phase I) costs much more time resources than the channel probing in Phase II, thus the

threshold for further channel probing in Phase I (i.e., θ_{L-1}^*) should be the smaller than

the thresholds in Phase II (i.e., $\{\theta_j^*, j = 0, 1, \ldots, L-2\}$).

4.2.2. Generalization To Heterogeneous Cases

In the following, we generalize the above study to the model where different transmit-

ters may have different numbers of receivers. Let L_m denote the number of receivers for

transmitter m. Without loss of generality, we assume that $L_1 \leq L_2 \leq \cdots \leq L_M$.

We have the following proposition regarding the optimal stopping rules in heterogenous networks using SPWR probing. We note that similar study can be carried over to SPWOR probing.

Proposition 4.2.4. *Suppose that sequential probing with recall strategy (SPWR) is used for channel probing. Then the optimal scheduling rule for distributed scheduling exhibits a multi-stage threshold structure with thresholds* $\{\theta_0^*, \theta_1^*, \ldots, \theta_{L_M-1}^*\}$. *Specifically, for the successful transmitter* $t(n)$, *the optimal thresholds are* $\{\theta_0^*, \theta_1^*, \ldots, \theta_{L_{t(n)}-1}^*\}$, *i.e., it continues probing until at some receiver* j $(0 \leq j \leq L_{t(n)} - 1)$, $R_{n,j} \geq \theta_j^*$, *followed by data transmission; otherwise, if* $R_{n,j} < \theta_j^*$ *for all* j, *then all links re-contend.*

Sketch of the proof. The proof is built upon Proposition 4.2.3, and hinges on the fact that the system restarts each time when a new channel contention begins. So we can use backward induction to derive the optimal stopping rule. Note that another key point is that the channel probing cost $x^*\tau$ is the same for all the transmitters. Therefore, for any transmitters $i, j \in \{1, 2, \ldots, M\}$, they share the same thresholds from receiver 0 to $\min(L_i - 1, L_j - 1)$.

We note that similar study can be carried over to SPWOR probing.

4.2.3. Channel Aware Distributed Scheduling for Multicast Traffic

In this section, we consider channel aware distributed scheduling for multicast traffic (see Fig. 4.5), where all receivers corresponding to one transmitter require the same data from that transmitter. In this case, the channel probing process follows the same line of that in the ESPWR mechanism for the unicast traffic (see Section 4.2.1.2): The transmitter probes all receivers to observe $\{R_{n,t(n),j}, j = 0, 1, \ldots, L - 1\}$. Hence for each data

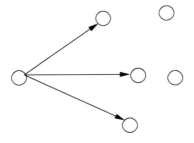

Figure 4.5. An example of multicast transmission.

transmission, the reward is a function of $\{R_{n,t(n),j}, j = 0, 1, \ldots, L - 1\}$. Depending on the reward of interest, we consider the following two multicast scenarios:

1. Case I: the reward is defined to be the number of ready receivers [31], where by "readiness" we mean the rate $R_{n,t(n),j}$ is larger than some threshold R_{th}. Therefore, the reward for each transmission is given by

$$R_1 = \sum_{j=0}^{L-1} \mathbf{I}\left[R_{n,t(n),j} \geq R_{th}\right]. \tag{4.27}$$

Clearly, R_1 is a binomial random variable with parameters (L, p), where $p = P(R_{n,t(n),j} \geq R_{th})$.

2. Case II: suppose the transmitter send the information at a constant rate R_c, and the reward is defined to be the sum rate, i.e.,

$$R_2 = R_c \sum_{j=0}^{L-1} \mathbf{I}\left[R_{n,t(n),j} \geq R_c\right]. \tag{4.28}$$

It is clear that R_2 is a product of a constant R_c and a binomial random variable with parameters (L, p), where $p = P(R_{n,t(n),j} \geq R_c)$.

For both scenarios, the optimal scheduling policy and the optimal throughput can be obtained as follows.

Proposition 4.2.5. *a) The optimal stopping rules for distributed scheduling in both multicast cases take the following form:*

$$N^* = \min\{n \geq 1 : R \geq x^*\}, \tag{4.29}$$

b) the maximum multicast network throughput x^* is a optimal threshold, and is the unique solution to

$$E(R - x)^+ = \frac{x[1 + p_s(L - 1)]\delta}{p_s}, \tag{4.30}$$

where R is a random variable with the same distribution as R_1 (or R_2).

4.3. Iterative Algorithm for Computing the Optimal Thresholds

4.3.1. Iterative Algorithm Using Fractional Maximization

In the following, we develop iterative algorithms for computing the thresholds.

First, observe that for both SPWOR and SPWR, the optimal stopping rules for distributed scheduling are multi-threshold policies. Denote the thresholds by $\vec{\theta} = [\theta_0, \theta_2, \ldots, \theta_{L-1}]^T$, and the average throughput by $\Phi(\vec{\theta})$. Based on (2.3), $\Phi(\vec{\theta})$ usually takes the form of $U(\vec{\theta})/V(\vec{\theta})$, and the optimal threshold $\vec{\theta}^*$ is then the one that maximizes $U(\vec{\theta})/V(\vec{\theta})$. However, since direct maximization of $U(\vec{\theta})/V(\vec{\theta})$ is often prohibitive, we resort to the technique of fractional maximization [13].

To this end, define the function $W(\vec{\theta}, x) \triangleq U(\vec{\theta}) - xV(\vec{\theta})$, where x is a real positive value. For a given x, the corresponding $\vec{\theta}(x)$ that maximizes $W(\vec{\theta}, x)$ is denoted as $\vec{\theta}(x) = \arg\max_{\vec{\theta}} W(\vec{\theta}, x)$.

Let $W(x) \triangleq W(\vec{\theta}(x), x)$, and x^* denote the solution to $W(x) = 0$. It can then be shown that x^* is the optimal throughput, and that $\vec{\theta}(x^*)$ is the optimal threshold. To construct iterative algorithms for computing x^* and $\vec{\theta}(x^*)$, we need the following lemma [39].

Lemma 4.3.1. $W(x)$ *is decreasing and convex in* x.

It can also be shown that for any given x', $z = W(x') - V(\vec{\theta}(x'))(x - x')$ is a supporting hyperplane for $W(x)$ at x'. Then, Newton's method yields that

$$x_{n+1} = x_n - \frac{W(x_n)}{W'(x_n)} = \Phi(\vec{\theta}(x_n)). \tag{4.31}$$

It can be shown that the above iterative algorithm converges quadratically to x^* [13]. Summarizing, we have the following proposition.

Proposition 4.3.6. *Given any positive initial value* x_0, *the following iterative algorithm*

$$\begin{cases} \vec{\theta}(x_n) = \arg\max_{\vec{\theta}} W(\vec{\theta}, x_n), \\ x_{n+1} = \Phi(\vec{\theta}(x_n)), \end{cases}$$

converges, i.e., $x_n \to x^*$ *and* $\vec{\theta}(x_n) \to \vec{\theta}^*$, *and the convergence rate is quadratic.*

Building on Proposition 4.3.6, in the following, we derive iterative algorithms for computing the optimal thresholds for the four probing mechanisms. For ease of exposition, we consider the continuous rate case only, i.e., we assume that the pdf $f(r) > 0$ for all $r > 0$. Similar studies can be carried out for the discrete rate case also.

4.3.2. Iterative algorithm for RS Probing and ESPWR Probing

Recall that the optimal scheduling algorithm for the RS mechanism is the same as that in the single-receiver case, i.e., a single-threshold policy. For a given threshold θ, the throughput of the RS mechanism can be shown to be

$$\Phi(\theta) = \frac{\int_\theta^\infty r \, dF(r)}{\delta/p_s + \int_\theta^\infty dF(r)}. \tag{4.32}$$

Accordingly, $U(\theta) = \int_\theta^\infty r dF(r)$ and $V(\theta) = \delta/p_s + \int_\theta^\infty dF(r)$. It can also be shown that (cf. Lemma 3.1 in [102]), for any given x_n, $\theta(x_n) = \arg\max_\theta W(\theta, x_n) = x_n$.

Thus, appealing to Proposition 4.3.6, for any positive initial value x_0, the iterates generated by the following algorithm:

$$x_{n+1} = \Phi(x_n), \tag{4.33}$$

converge to the optimal threshold and the maximum network throughput x^* quadratically.

We note that the same algorithm in (4.33) can be applied to the ESPWR mechanism as well.

4.3.3. Iterative algorithm for SPWOR Probing

Given a threshold $\vec{\theta}$, it can be shown that the average throughput of the SPWOR mechanism is given by

$$\Phi(\vec{\theta}) = \frac{p_s \left[\sum_{j=0}^{L-1} p_j \frac{\int_{\theta_j}^\infty r dF(r)}{1 - F(\theta_j)} T \right]}{(1 - p_s)\tau + p_s \left[\left(1 - \sum_{j=0}^{L-1} p_j \right) L\tau + \sum_{j=0}^{L-1} p_j((j+1)\tau + T) \right]}, \tag{4.34}$$

where $p_j = \prod_{i=0}^{j-1} F(\theta_i)(1 - F(\theta_j))$ is the probability that the transmitter transmits to the j-th receiver.

It is not difficult to show that

$$\theta_{L-1} = x, \tag{4.35}$$

and for $j = L - 2, L - 3, \dots, 0$,

$$\begin{aligned}
\theta_j &= \sum_{k=j+1}^{L-1} \prod_{i=j+1}^{k-1} F(\theta_i) \int_{\theta_k}^\infty r dF(r) \\
&- x \left\{ \sum_{k=j+1}^{L-1} \prod_{i=j+1}^{k-1} F(\theta_i)(1 - F(\theta_k)) \left[1 - (L - 1 - k)\frac{\tau}{T} \right] + (L - 1 - j)\frac{\tau}{T} - 1 \right\}
\end{aligned} \tag{4.36}$$

are the optimal thresholds that maximize $W(\vec{\theta}, x)$.

We note that the updating procedure of these thresholds follows a backward order from $L - 1$ to 0, since the updating of θ_j (using (4.36)) requires the knowledge of the updated values of θ_{j+1} to θ_{L-1}.

Summarizing, for every x_n, the corresponding $\vec{\theta}(x_n)$ can be obtained using (4.35) and (4.36), and then x_{n+1} can be found using $x_{n+1} = \Phi(\vec{\theta}(x_n))$ in (4.31).

4.3.4. Iterative algorithm for SPWR Probing

According to Proposition 4.2.2, without loss of optimality, we can assume that $\theta_{L-1} \leq \theta_0 \leq \theta_1 \leq \cdots \leq \theta_{L-2}$. It can then be shown that the average network throughput of the SPWR mechanism is given by (4.37).

$$\Phi(\vec{\theta}) = \frac{\sum_{j=0}^{L-2} p_j \frac{\int_{\theta_j}^{\infty} rf(r)dr}{1-F(\theta_j)} + \frac{\frac{\int_{\theta_{L-1}}^{\theta_0} Lrf(r)(F(r))^{L-1}dr}{\prod_{j=1}^{L-2} F(\theta_j)} + \sum_{j=0}^{L-3} \frac{\int_{\theta_j}^{\theta_{j+1}}(L-j-1)rf(r)(F(r))^{L-j-2}dr}{\prod_{i=j+1}^{L-2} F(\theta_i)} + \int_{\theta_{L-2}}^{\infty} rf(r)dr}{\frac{1}{p_{L-1}}\left[1 - \frac{(F(\theta_{L-1}))^L}{\prod_{j=0}^{L-2} F(\theta_j)}\right]}}{(1/p_s - 1)\delta + \left[\left(1 - \sum_{j=0}^{L-1} p_j\right)L\delta + \sum_{j=0}^{L-1} p_j((j+1)\delta+1)\right]}. \qquad (4.37)$$

Note that p_j, the probability that the transmitter transmits to the j-th receiver, is now given by

$$p_j = \prod_{i=0}^{j-1} F(\theta_i)(1 - F(\theta_j)), \ \forall \ j = 0, 1, \ldots, L - 2, \qquad (4.38)$$

$$p_{L-1} = \prod_{i=0}^{L-2} F(\theta_i)\left[1 - \frac{(F(\theta_{L-1}))^L}{\prod_{j=1}^{L-2} F(\theta_j)}\right]. \qquad (4.39)$$

In practice, the number of receivers per transmitter is usually not large. We also note that most multi-receiver gain occurs when $L = 2$ and $L = 3$. In what follows, we study the

case when $L = 2$. It can be shown that the following two time-scale algorithm converges
to the optimal thresholds of SPWR:

$$\vec{\theta}(x_n) = [x_n, y^*(x_n)]^T, \tag{4.40}$$

$$x_{n+1} = \Phi(\vec{\theta}(x_n)), \tag{4.41}$$

where $y^*(x_n)$ is the limit of the following iterative algorithm:

$$y_{m+1} = \frac{\int_{y_m}^{\infty} r dF(r) - x_n \delta}{1 - F(y_m)}, \tag{4.42}$$

for any positive initial value y_0.

4.4. Numerical Results

In this section, we provide numerical results for the continuous rate case, assuming
that the transmission rate is given by the Shannon channel capacity:

$$R(h) = \log(1 + \rho h) \text{ nats/s/Hz},$$

where ρ is the normalized average SNR, and h is the random channel gain corresponding
to Rayleigh fading. Therefore, the transmission rate R has the following distribution

$$F(r) = 1 - \exp(-\frac{\exp(r) - 1}{\rho}). \tag{4.43}$$

Unless otherwise specified, we will fix $p_s = \exp(-1)$.

We first examine the convergence rate of the iterative algorithm proposed in Section 4.3, and specifically, the iterative algorithm (4.35), (4.36) and (4.31) for the SPWOR
mechanism. The results are presented in Table 4.1 and Table 4.2. Specifically, Table 4.1
presents the convergence of x_n for different ρ and δ. It can be observed that the iterative

Table 4.1. Convergence of the iterative algorithm for SPWOR ($L = 3$).

(ρ, δ)	x_0	x_1	x_2	x_3	x_4
(0.5, 0.1)	0.5	0.4511	0.4521	0.4521	0.4521
(0.5, 0.5)	2	0.1533	0.1966	0.1969	0.1969
(1, 1)	0.5	0.1740	0.1921	0.1922	0.1922
(10, 1)	1	1.9567	1.9889	1.9890	1.9890

Table 4.2. Convergence of the thresholds ($L = 3, \rho = 1, \delta = 1$).

$Iterations$	0	1	2	3
θ_0	0.1201	0.5164	0.4921	0.4920
θ_1	0.2185	0.4374	0.4226	0.4225
θ_2	0.5000	0.1740	0.1921	0.1922

algorithm converges fast (within 3 or 4 iterations). Table 4.2 presents the convergence

behavior of the thresholds for given ρ and δ.

Next, we compare the performance of SPWOR, ESPWR and RS probings. Fig.4.6

depicts the throughput of these three strategies as the number of receivers increases. It

can be seen that for the same setting, the throughput of the SPWOR mechanism is always

the largest. Note that although the throughput of SPWOR increases as L increases, the

performance gain saturates, and it can be observed that most multi-receiver gain occurs for

$L = 2$ and $L = 3$. We also observe that there exists an optimal L^* where the throughput

gain of ESPWR over RS is maximum. For example, $L^* = 2$ for $\rho = 0.5$ and $\delta = 0.5$.

We also compare the performance between SPWOR and SPWR when $L = 2$. The

results are presented in Table 4.3. Clearly, the throughput of SPWR is always higher

than that of SPWOR, but their performance is comparable. Given the complexity of the

SPWR mechanism, we argue that the SPWOR mechanism is more preferable in practice.

Last, we study the impact of the parameters ρ and δ on the performance of the optimal

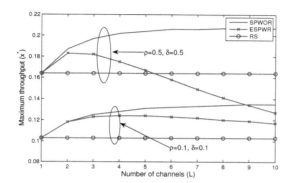

Figure 4.6. Throughput as a function of number of receivers.

Table 4.3. Performance comparison between SPWR and SPWOR ($L = 2$).

(ρ, δ)	$(0.1, 0.1)$	$(0.5, 0.5)$	$(0.5, 1)$	$(1, 1)$
x^*_{SPWR}	0.119	0.190	0.116	0.187
x^*_{SPWOR}	0.118	0.187	0.114	0.185

scheduling. Using SPWOR as an example, Table 4.4 outlines the performance gain of SPWOR as a function of L, comparing to RS probing. As expected, the performance gain decreases as ρ increases. However, Table 4.4 reveals that the performance gain can be neither decreasing nor increasing in δ when the number of receivers is small. Indeed, for $L = 2$, it can be observed that the performance gain first increases as δ decreases from 1 to 0.5, and then decreases as δ further decreases to 0.1. Our intuition is that the multiuser diversity gain dominates the multi-receiver diversity gain when δ is small.

Table 4.4. Performance gain as a function of L.

L	2	3	4	5
$\rho = 0.5, \delta = 1$	13.97%	19.35%	21.82%	23.04%
$\rho = 0.5, \delta = 0.5$	14.32%	20.14%	23.00%	24.54%
$\rho = 0.5, \delta = 0.1$	12.28%	17.64%	20.61%	22.45%
$\rho = 1, \delta = 1$	12.62%	17.22%	19.21%	20.14%
$\rho = 1, \delta = 0.5$	12.98%	18.02%	20.40%	21.61%
$\rho = 1, \delta = 0.1$	11.08%	15.81%	18.38%	19.95%
$\rho = 10, \delta = 1$	7.24%	9.16%	9.76%	9.96%
$\rho = 10, \delta = 0.5$	7.79%	10.16%	11.00%	11.33%
$\rho = 10, \delta = 0.1$	7.16%	9.95%	11.32%	12.08%

4.5. Conclusion

In this chapter, we studied channel aware distributed scheduling for ad hoc communications where each transmitter node has multiple intended receivers. A key observation is that channel probing must be done effectively to balance the tradeoff between the throughput gain from better channel conditions and the probing cost. We characterized the optimal tradeoff in a stochastic decision-making framework, and developed the corresponding distributed scheduling to leverage the multi-receiver diversity and multiuser diversity in a joint manner. Specifically, we studied the scheduling problem using optimal stopping theory, and considered four probing mechanisms in phase II, namely, 1) random selection, 2) exhaustive sequential probing with recall, 3) sequential probing without recall, and 4) sequential probing with recall. We derived the corresponding optimal scheduling policies. We showed that these optimal scheduling policies have threshold structures, and the optimal thresholds can be obtained off-line by solving fixed point equations. Therefore, the optimal scheduling policies are amenable to easy distributed implementation. We further devised iterative algorithms to obtain the optimal thresholds. We established the convergence of the iterative algorithm, and showed that the convergence speed is quadratic. Numerical examples were provided to corroborate the theoretic findings.

CHAPTER 5

PROTOCOL DESIGN AND THROUGHPUT ANALYSIS OF FREQUENCY-AGILE

MULTI-CHANNEL MEDIUM ACCESS CONTROL

5.1. Introduction

In the previous chapters, we have studied distributed opportunistic scheduling under a stochastic decision-making framework. In this chapter, we turn out attention to
protocol design of MAC to exploit channel diversity for multi-channel wireless ad hoc
networks [103].

Our main contributions in this chapter are two folds. First, we take channel fading
into consideration in the MAC protocol design. We propose an opportunistic multi-
channel MAC protocol (OMC-MAC), which in fact makes use of the channel variations
across multiple frequency channels to boost up the system throughput; and it is in this
sense that we call the proposed scheme "opportunistic multi-channel MAC". In particular,
we devise the OMC-MAC based on the IEEE 802.11 distributed coordination function
(DCF) by using two mechanisms provided by the 802.11 standard. The first mechanism
is the request-to-send (RTS) and clear-to-send (CTS) handshake. In the OMC-MAC
protocol, the RTS/CTS handshake is used not only to reserve the transmission floor
and select the channel, but also as pilot signaling to measure the channel condition for
rate adaptation. Note that different from Chapter 4, we assume that the channel state
information for all the available channels can be obtained in one RTS-CTS handshake.
Hence, by selecting the channel with the best condition, users can send packets at higher
rates. The corresponding multi-channel gain can be viewed as *selection diversity gain*.

[1]©[2009] IEEE. Reprinted, with permission, from Dong Zheng and Junshan Zhang, "Protocol Design and Performance Analysis of Frequency-Agile Multi-Channel Medium Access Control", IEEE Trans. on Wireless Communications, Vol. 5, Issue 10, Oct. 2006 Page(s):2887-2895.

The second mechanism is the PHY-layer multi-rate capability, enabling that transmissions can take place at multiple rates according to the channel conditions. For example, in the IEEE 802.11b, the data rates can be 2, 5.5, and 11Mbps, and the data rates of 802.11g can be as high as 54Mbps. We note that the operation of OMC-MAC requires nodes to have multi-band communication capabilities, which can be made possible by using cognitive radio techniques.

To characterize the throughput gain of the proposed multi-channel MAC, we study the saturation throughput defined as the throughput in heavy traffic conditions, under which each node always has packets to transmit [18]. We derive the throughput gain per channel usage of OMC-MAC over the single-channel MAC in WLANs. The analysis results reveal that the OMC-MAC protocol achieves significant multi-channel diversity gain, even under heavy traffic conditions.

Our second contribution is the use of the particle filter to estimate the number of competing stations, and adjust the size of the contention window accordingly. Recent works on the IEEE 802.11 standard have shown that the MAC performance hinges heavily on the number of competing stations [18, 89]. It is sensible to adapt the size of the contention window to the number of competing stations so as to optimize the performance (see, e.g., [19, 27]). To this end, an important step is to estimate the number of the competing stations. Notably, two techniques, namely ARMA filtering and extended Kalman filtering (EKF), have been proposed for this estimation purpose, which is based on the relationship between the number of competing stations and the collision probability in saturation conditions [19, 20]. Observing that this mapping is non-linear in nature, we propose to use the *particle filtering* method instead. Generally speaking, the particle filtering technique is a sequential Monte Carlo (SMC) method, and is particularly useful for non-linear/non-Gaussian tracking problems. The basic idea of this method is to use random samples ("particles") to represent the PDF, rather than as a function of the state space. When the number of the particles is reasonably large, they can provide a good approximation of the probability model under consideration. As a result, statistical moments, such as mean and variance, can be computed directly from these samples.

We have carried out extensive simulation experiments to evaluate the OMC-MAC performance compared with that of the standard 802.11 DCF and other multichannel protocols such as DCA and MMAC [95, 84]. The simulation results reveal that OMC-MAC can achieve efficient channel utilization for each added channel. Furthermore, we integrate the particle filtering algorithm into OMC-MAC to adapt the contention window size. The simulation results show that this enhancement yields a throughput gain of 12%.

The rest of the chpater is organized as follows. Section 5.2 provides background and related work. The detailed description of the proposed OMC-MAC is given in Section 5.3, followed by the analysis of the saturation throughput in Section 5.4. Section 5.5 discusses the simulation settings and analyzes the simulation results of OMC-MAC. Section 5.6 presents the particle filtering approach to the estimation of the number of competing stations. Finally, Section 5.7 concludes this chapter and discusses future work.

5.2. Background and Related Work

In related work, recent studies [95, 84, 70, 46] have introduced multi-channel MAC protocols for ad hoc networks. Different from single-channel MAC, a multi-channel MAC protocol involves channel allocation. Recent work [70] assumes that nodes can listen to all channels concurrently, which requires multi-transceivers at each host. By building a free-channel list through sensing, a node can transmit after randomly choosing one from the list. In contrast, [84] proposes to use only one transceiver per host, and the channel reservation is done through the use of the Ad hoc Traffic Indication Messages (ATIM) window. A Dynamic Channel Assignment (DCA) protocol is proposed in [95] where the channels are assigned in an on-demand fashion. Different from the two algorithms noted above, one channel is dedicated solely for the control purpose. It is proposed in [46] that the data channels can be separated from the control channel, and the channel selection is decided at the receiver side by choosing the data channel as the one with the least interference.

Along a different line, there have recently been studies on exploiting channel variations for rate control in ad hoc networks. It is shown in [44] that by changing the modulation scheme and hence the transmission rate corresponding to different channel conditions, there can be a dramatic increase in the bandwidth efficiency. Their proposed rate adaptive MAC protocol is named as the Receiver-Based AutoRate (RBAR) protocol. [81] extends RBAR by opportunistically sending multiple back-to-back data packets whenever the channel condition is good, and this scheme is called the opportunistic auto rate (OAR) protocol. It will be made clear that one key feature distinguishing the proposed OMC-MAC from RBAR and OAR is that the OMC-MAC scheme "aggressively" seeks the best channel condition across multiple channels to achieve diversity gain, whereas the latter adapts the rate accordingly to the given channel conditions.

5.3. The OMC-MAC Protocol Description

The proposed OMC-MAC protocol is built on the single-channel 802.11 DCF. In particular, we assume that nodes can listen to multiple channels simultaneously by using a programmable-cognitive radio, but can be engaged in one data transmission for the sake of fairness.

Suppose that there are totally L orthogonal frequency channels available, among which M channels can be utilized for data transmissions in the multi-channel system. We assume that these M channels can be any combinations out of the L channels. Note that such a limitation on the channel usage is needed for sharing the unlicensed industrial, scientific, and medical (ISM) band, so that other devices using the same ISM band can work as well. Furthermore, we will show in Section 5.4 that by doing this, a diversity gain of order $L - M + 1$ can be achieved, even in heavy traffic conditions.

In OMC-MAC, every node is required to maintain a busy channel list (BCL) and a free channel list (FCL), which are updated by the control packets. The i_{th} entry of the BCL for node A, denoted as $A.BCL[i]$, has four fields (<*source, destination, channel_ID, release_time*>), where the *channel_ID* indicates which channel the transmission takes place and the *release_time* tells when the transmission finishes. Via using BCL, at each time slot, a node can tell how many channels are being used, which channels are free and which neighboring nodes are busy in transmissions. The FCL can be easily generated from the BCL. Then nodes can use the FCL to initiate the transmissions.

The basic idea of OMC-MAC can be outlined as follows. When a node recognizes that there are free channels available, it broadcasts RTS over all free channels listed in the FCL, using the basic rate. As a result, all neighboring idle nodes would hear that RTS packet. Once the intended receiver receives RTS over the free channels, it computes the signal to noise ratio (SNR) obtained from these RTS signals and chooses the channel with the highest SNR as the desired one. After that, the receiver transmits the CTS packet in the chosen channel using the basic rate. Note that the rate for data transmissions is also piggybacked with CTS. The rest of the MAC protocol can be performed using the same mechanism as in the RBAR or OAR protocols, i.e., the data transmissions use the selected channel and the transmission rate.

In what follows, we outline some challenges for general multi-channel MAC protocols, along with the solutions provided by OMC-MAC. Assume that node A and node B are in the communication range of each other; however, node D cannot hear B's transmission and C cannot hear A's. We have the following observations.

Ambiguity of channel selection and reservation time: Suppose that A sends RTS to B through all free channels in the FCL. D hears the RTS packet and updates its BCL accordingly. Note that this is a tentative reservation [44], and this reservation only prevents D from transmitting. That is to say, it may not reveal the actual transmission duration and the channel, and the BCL shall be updated after the channel and transmission rate are selected. After B receives RTS, it picks the channel with the highest SNR and sends back CTS over this channel with a selected rate. The nodes around B, such as C, update their BCLs accordingly. However, since D does not hear B's CTS, D cannot determine which channel A would use and the actual transmission duration therein. This is because the subsequent data transmission between A and B would use the chosen rate, which may be different from the basic rate. Hence, D may not decode the data packet header correctly, making the BCL of D outdated.

In order to resolve the above ambiguity issues while keeping the PHY-layer intact, we propose to add an RES packet for reservation to the RTS/CTS handshake. The RES frame format is shown in Fig. 5.1. The *frame control* field contains the information of frame type. The *rate* field carries the tx/rx rate of data packets. The *packet length* field gives the size of the data packet. The *FCS* is the frame check sequence which protects the frame. Note that in OMC-MAC, the RES and CTS control frames have the same format as the RTS frame.

2 Octets	4 Bits	12 Bits	6 Octets	6 Octets	4 Octets
Frame Control	Rate	Packet Length	Destination Address	Source Address	FCS

Figure 5.1. The RTS/CTS/RES frame format of OMC-MAC

Simply put, after receiving CTS from B, A uses the basic rate to send an RES packet which includes the source-destination pair, the packet length and the rate over the selected channel. Hence, D can decode this packet correctly and update its BCL accordingly. In fact, RES is used not only to inform the selected channel, but also help to update the entries of the BCL lists of other nodes. Recall that the tentative reservation of D needs to be updated using the right information such as channel ID, transmission rate and duration. This happens when the rate chosen by the receiver is different from the basic rate or if there is more than one free channels. In the first case, the value of the *release_time* obtained from RTS is not valid, and the new value should be calculated by using the transmission rate and the packet length carried by the RES packet. In the later case, RES is needed to "clean" the corresponding entries of the BCL because all other free channels which had been marked as busy (due to RTS) are still free, except the chosen one. In order to make the above updates, the information on the *<source, destination>*

pair in the RES packet is needed to update the BCL structure.

We note that the <*source, destination*> pair in the BCL can also be used to prevent nodes from sending RTS blindly. Recall that in the 802.11 DCF, only the transmission time is recorded and there is no need for recording the transmission source or destination. This is because in a single channel environment, no other nodes would transmit if the channel is busy, whereas it is different in a multi-channel environment. Suppose that D does not record the source or destination addresses, and later on, D wants to transmit to A. If there is a free channel available, D would start to transmit to A. Since A is busy, A cannot respond to that RTS with CTS. Therefore, D would time out for the expected CTS packet. After trying many times, the link layer of D would report to the network layer that the path to A is unavailable which subsequently causes un-necessary re-routing. Therefore, with <*source, destination*> recorded in the BCL, D could be prevented from sending RTS blindly.

For the same reason noted above, we add the <*source, destination*> information into the CTS frame so as the nodes in the vicinity of the sender/receiver could update their BCLs accordingly.

The hidden terminal problem: Consider a scenario where B is sending data frames to C using channel i. At this time, D starts a transmission to A using a different channel j. If B cannot receive the CTS packet from A, then B's BCL is outdated. Suppose later on B finishes its data transmission and attempts another transmission. Since B cannot sense the transmission from D, and furthermore channel j is in the FCL, it may start transmissions on j, which would interrupt the data reception at A. This is a typical hidden terminal problem in multi-channel MAC, although it happens with low probability due to

96

the contention. This hidden terminal problem can be resolved if B is enabled to intercept exchanges of control packets on the other channels during its own data transmissions, for example, using cognitive radio techniques.

To get a more concrete sense of the operation of OMC-MAC, we now give an example (see Fig. 5.2).

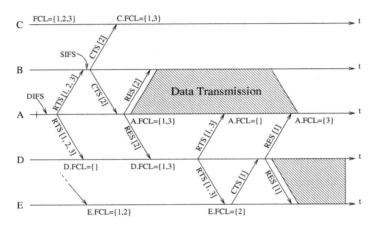

Figure 5.2. Operation of the OMC-MAC protocol (the numbers inside the bracket indicate the channel ID)

Example 1: Suppose that A wants to transmit to B. The basic steps of the proposed OMC-MAC scheme are outlined as follows:

1. Node A checks the following conditions: 1) node B is one of the hosts that are active in transmissions, i.e., B is equal to $A.BCL[i].source$ or $A.BCL[i].destination$ for some i; 2) the FCL is empty; and 3) there are M transmissions in the BCL list. If any of the above conditions is true, node A would refrain from transmitting the RTS and wait until all the conditions become false. This is the virtual carrier sensing.

2. Before transmissions, A senses the channels in the FCL list for a distributed interframe spacing (DIFS) time and waits for a random backoff time. After that, if none of the channels in the FCL are available, or there are already M transmissions, go back to step **1**. Otherwise, node A sends RTS through all remaining free channels in the FCL and waits for CTS. If node A does not receive CTS in a time-out period, A assumes that a collision has occurred and would use the binary exponential backoff algorithm (BEB) of the 802.11 DCF. Note that the adaptive contention window algorithm will be applied after the particle filtering technique is introduced in Section 5.6.

3. If B correctly receives RTS through all the free channels, it calculates the SNR from the RTS and selects the channel with the highest SNR. It then piggybacks the transmission rate and the packet length within the CTS packet, and sends it back via the selected channel. However, if the intersection of $A.FCL$ and $B.FCL$ is empty, B would not send the CTS packet since it would cause collision with another ongoing transmission.

4. Node A receives the CTS packet and transmits the data packets using the designated transmission rate and the selected channel. Before the data transmission, node A sends an RES packet with the packet length and the transmission rate through the selected channel to prevent other nodes from using this channel.

5. Other nodes, upon receiving the CTS and RES packets, update the BCL and FCL correspondingly.

Note that the FCL is built from the BCL at the time when a node is trying to reserve the channel. Recall that a node has to wait for a DIFS period plus a random backoff before accessing the channels. Hence, it is possible that one or more of the free channels become busy and are occupied by other nodes during this period. These busy channels would be deleted from the FCL immediately upon detection. Unless the FCL becomes empty, or there are already M busy channels, the node would continue to wait for the remaining time of the DIFS period and the random backoff. Furthermore, some channels in the busy channel list may become free during this period. However, these channels would not be added to the free channel list for the current waiting period. In a word, the number of free channels would decrease during the DIFS and the random backoff period. After this period, if free channels are still available, and the number of busy channels are less than M, RTS packets will be sent out through these remaining free channels.

5.4. Analysis of Saturation Throughput

In this section, we study the throughput gain per channel usage of OMC-MAC over the single-channel MAC in WLANs. Let n_1 and n_M denote the numbers of nodes in the single-channel system and the multi-channel system. We assume that the adaptive contention window mechanism is applied, and that there is a one-to-one mapping between the contention window size W and the number of competing nodes, i.e., $W_1 = n_1 K$ and $W_M = n_M K$, where K is a constant [18].

We assume that different channels experience independent Rayleigh fading, and the channel statistics are the same across all channels between two nodes. We assume for now that the transmission rate is proportional to the SNR, and the transmission duration is constant.

Define

$$\alpha \triangleq \frac{(1 - \frac{2}{n_M K})^{n_M - 1}}{(1 - \frac{2}{n_1 K})^{n_1 - 1}}. \tag{5.1}$$

It will be made clear below that this value is the ratio of the successful transmission probability of the multi-channel MAC to that of the single-channel MAC. We have the following proposition.

Proposition 5.4.1. *Given F free channels, the saturation throughput gain per channel usage of the opportunistic multi-channel MAC (OMC-MAC) over the single-channel MAC, denoted as Γ, is*

$$\Gamma = \alpha \sum_{j=1}^{F} \frac{1}{j} - 1. \tag{5.2}$$

Proof: Recall that in OMC-MAC, nodes contend for the channel by sending RTS through all the free channels in the FCL. Therefore, the contention process in the proposed multi-channel MAC is essentially the same as that in the 802.11 DCF for a single channel system. It follows that the same Markov model in [18] is applicable here. Since the number of backoff stages m is 0, the probability that a station transmits in a randomly chosen time slot is [18]

$$\tau = \frac{2}{W+1}, \tag{5.3}$$

and the successful transmission probability can be calculated as

$$P_s = n\tau(1 - \tau)^{n-1}. \tag{5.4}$$

Appealing to [18], the average throughput gain per channel usage of the multi-channel system over the single-channel system can be shown to be

$$\Gamma = \frac{n_M \tau_M (1 - \tau_M)^{n_M-1} \times \mathbb{E}[\text{SNR}_{max}(F)]}{n_1 \tau_1 (1 - \tau_1)^{n_1-1} \times \mathbb{E}[\text{SNR}_1]} - 1, \tag{5.5}$$

where τ_M and τ_1 denote the transmission probabilities of the multi-channel system and the single-channel system.

It is easy to see that the SNR of channel k, denoted as $\gamma_k = \rho|h_k|^2$, for $k = 1, ..., F$, is exponentially distributed with mean $\bar{\gamma} = \rho$, where ρ is the average SNR at the receiver,

100

and h_k is a complex-Gaussian random variable with zero-mean and unit-variance. It follows that the mean of $\text{SNR}_{max}(F)$ is given by [97]

$$\mathbb{E}[\text{SNR}_{max}(F)] = \rho \sum_{j=1}^{F} \frac{1}{j}. \tag{5.6}$$

Hence, the multi-channel diversity gain is

$$G(F) = \frac{\mathbb{E}[\text{SNR}_{max}(F)]}{\mathbb{E}[\text{SNR}_1]} = \sum_{j=1}^{F} \frac{1}{j}. \tag{5.7}$$

Recall that $\tau \approx \frac{2}{W}$ when W is large. Since $W_M = n_M K$ and $W_1 = n_1 K$, we conclude that

$$\Gamma = \frac{(1 - \frac{2}{n_M K})^{n_M - 1}}{(1 - \frac{2}{n_1 K})^{n_1 - 1}} G(F) - 1 = \alpha G(F) - 1. \tag{5.8}$$

$\blacksquare$

Remarks: We note that it is difficult to characterize the dynamics of the number of free channels F under general traffic conditions. Instead, we consider the interesting heavy traffic regime, i.e., there are many nodes in both the single-channel and multi-channel systems ($n_M \gg 1$ and $n_1 \gg 1$). Using Proposition 5.4.1, we have the following result:

$$\lim_{\substack{n_1 \to \infty \\ n_M \to \infty}} \Gamma = \lim_{\substack{n_1 \to \infty \\ n_M \to \infty}} \frac{(1 - \frac{2}{n_M K})^{n_M - 1}}{(1 - \frac{2}{n_1 K})^{n_1 - 1}} G(F) - 1 = G(F) - 1, \tag{5.9}$$

where $F \to L - M + 1$ since all the M channels are occupied when the traffic load is high.

The above result is based on the assumption that the rate is proportional to the SNR. For other rate functions $R(SNR)$, the throughput gain can be characterized along the same line. To get a concrete sense of the OMC-MAC performance, we plot in Fig. 5.3 the ratio T of the average throughput of the multi-channel system to that of L single-channel systems, for rate functions $R_1(x) = x$ and $R_2(x) = \log_2(1+x)$. The results reveal that the throughput performance of OMC-MAC benefits from the diversity gain significantly when F is 2 or 3. We also note that for the rate function R_2, the performance gain decreases as the average SNR ρ increases.

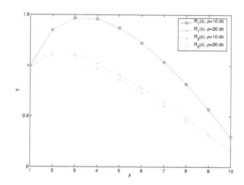

Figure 5.3. Normalized throughput vs. F $(L = 10)$

5.5. Simulation Experiments

In this section, we use simulations to evaluate the performance of OMC-MAC. Specifically, we use the GloMoSim simulator [10] to implement and evaluate the OMC-MAC protocol. To this end, we have made necessary modifications of some structures and functions in the original GloMoSim code to incorporate the cross-layer interactions. We have implemented OMC-MAC based on the RBAR protocol. We assume that the available rates are 2 Mbps, 5.5Mbps, and 11 Mbps. Following RBAR, the different rates are selected based on the corresponding SNR threshold and the instantaneous SNR [44]. Note that these rates are used to transmit data frames only, and the transmissions of the control packets use the basic rate, 2 Mbps. For fair comparison, OMC-MAC uses the same contention method as that of the standard 802.11 DCF, i.e., the BEB algorithm.

Unless otherwise specified, we assume that $L = M = 3$, i.e., there are totally 3 channels. As noted before, we assume independent Rayleigh fading across the channels. Parameters, such as slot time, DIFS period, radio transmitted power are defaulted values in GloMoSim [10]

The nodes are uniformly distributed in a designated area and are stationary. Half of the nodes are chosen randomly transmitting to the other half. Constant Bit Rate (CBR) traffic is generated between any pair of transmitters and receivers. The packet arrival rate can be adjusted to vary the network load. Unless otherwise specified, the packet size is fixed to be 1460 Bytes. In the network layer, we choose AODV as the routing protocol.

We study the performance of the OMC-MAC protocol in WLANs and multi-hop networks respectively, via varying some important network factors, such as offered traffic load (including the number of flows and packet inter-arrival time), data packet size and number of channels. Note that the *aggregate throughput* is used as the performance metric, which is obtained by summing over all achieved throughputs of the flows. The simulation results are gathered based on 20 replications of a 20-minute simulation.

5.5.1. Simulation Results

5.5.1.1. Impact of Traffic Load

In this part, we examine the impact of the offered traffic load on the protocol performance by changing the number of flows and the packet inter-arrival time. The simulation results for WLANs are shown in Fig. 5.4. As expected, when the network load is low, OMC-MAC does not bring much advantage over the single channel counterpart, simply because when the inter-arrival time of the packets is large, one channel is enough to handle all the traffic.

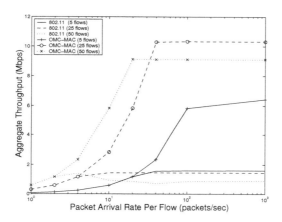

Figure 5.4. Aggregate throughput vs. packet arrival rate in WLANs

The gain of OMC-MAC over the single channel 802.11 DCF is clearly illustrated when the network load is sufficiently large. The aggregate throughput gain using three channels can be as high as 900%. The underlying rationale is as follows: 1) in OMC-MAC, the best channel in terms of SNR is chosen to conduct data transmissions; 2) rate adaption is used to exploit the best channel condition; and 3) concurrent transmissions are allowed because there are multiple channels. In contrast, recall that in the single channel MAC, there is no choice of the transmission medium (since there is only one channel), no matter the channel condition is good or bad. Furthermore, there is no rate adaptation in the basic 802.11 MAC protocol.

We also find that when the traffic load is high, the saturation throughput is reached for both single channel MAC and OMC-MAC, the main difference being that OMC-MAC enters the saturation region later than the single channel MAC. Hence, by grouping the available channels and use OMC-MAC, the stability is improved.

The simulation results for multi-hop networks are depicted in Fig. 5.5. As expected, given the same number of flows, the throughput of OMC-MAC and the single channel 802.11 DCF for multi-hop networks is smaller than that of WLANs (e.g., see Fig. 5.4 and Fig. 5.5). Despite the throughput loss, similar observations can be made on the performance gain of OMC-MAC over the standard 802.11 DCF. Another interesting behavior of the throughput performance in the multi-hop networks is that the saturation throughput in fact increases to a maximal point, then decreases slowly, which is perhaps due to the RTS/CTS-induced congestion [78].

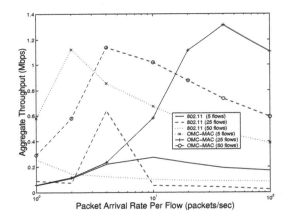

Figure 5.5. Aggregate throughput vs. packet arrival rate in multi-hop networks

5.5.1.2. Impact of Packet Size

Next we study the performance of MAC protocols by varying the data packet size. We plot in Fig. 5.6 the change of the aggregate throughput when the packet size ranges from 64 Bytes to 1024 Bytes. In general, we can find that the throughput performance of the network improves when the size of the data packet increases due to the decreased

overhead of the RTS/CTS(/RES) handshake for both OMC-MAC and the 802.11 DCF.

However, it is easy to see that the OMC-MAC performance increases much faster than

the 802.11 standard does. This is because good channel conditions have been utilized in

OMC-MAC. With a larger data packet, this utilization is even more efficient.

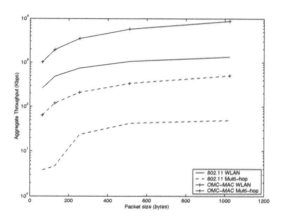

Figure 5.6. Aggregate throughput vs. packet size

Another interesting observation is that the performance improvement of OMC-MAC

in WLANs grows much faster as a function of the packet size, than that of multi-hop

networks, which is understandable since the channel conditions in WLANs are usually

much better than that of multi-hop networks (due to less interference).

5.5.1.3. Impact of Mobility

In this section, we investigate the effect of mobility on the performance of OMC-MAC.

To this end, we use the *random waypoint* model [10], in which a node moves according to

the following way: First a node randomly selects a destination from the physical terrain, and then it moves in the direction of the destination in a speed uniformly chosen between a minimum speed and a maximum speed. After it reaches its destination, the node stays there for a "pause-time" period. To study the effect of mobility on OMC-MAC, we fix the minimum speed as 0, vary the maximum speed from $1m/sec$ to $10m/sec$ and the pause-time from $1sec$ to $40sec$.

The simulation results are depicted in Fig. 5.7 for WLANs. Note that the mobility affects the coherence time of the fading channel and the propagation attenuation. A key observation can be drawn from the figure is that the maximum speed affects the throughput performance more than the pause-time. When the maximum speed is reduced from $10m/sec$ to $1m/sec$, the throughput gain increases by 8.5%. In contrast, the increase of the pause-time from $1sec$ to $40sec$ brings only a 2% throughput gain.

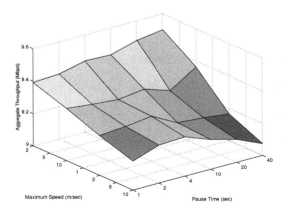

Figure 5.7. Aggregate throughput vs. mobility

5.5.1.4. Impact of Number of Channels

In this section, we examine the OMC-MAC performance against the number of channels in two different settings. First, let $L = M$ and increase them from 1 to 4. We investigate the channel utilization for every newly added channel in the OMC-MAC protocol and compare it with other multi-channel protocols. It should be mentioned that the OMC-MAC protocol boils down to the RBAR protocol when $L = 1$.

The simulation results for WLANs and multi-hop networks are shown in Table 5.1. Our results reveal that in WLANs there is a linear relationship between the aggregate throughput and the number of channels. The channel utilization is almost 100% for every added new channel (compared to the improvement of 80% per channel from MMAC and 60% from DCA in the same traffic regime [84]).

Table 5.1. OMC-MAC throughput vs. number of channels

L=M	WLANs (Mbps)	multi-hop networks (Kbps)
1	2.93	53.47
2	5.93	246.98
3	8.94	474.92
4	12.01	732.13

Table 5.1 shows that the channel utilization performance in multi-hop networks is much higher than that in WLANs. This is primarily because in multi-hop networks, spacial reuse "generates" more free channels, and hence results in higher channel diversity. In a nutshell, channel grouping achieves "resource pooling" in OMC-MAC and can bring a significant throughput gain.

Next, the simulations are performed in WLANs by fixing $L = 5$ and letting M increase from 2 to 5. Consequently, the diversity order F decreases from 4 to 1 under heavy traffic conditions. In order to compare the simulation results with the analysis in Section 5.4, we set the rates to be proportional to the SNR. We also apply the adaptive contention window scheme in the simulations. To examine the throughput improvement brought by the channel diversity, the performance of OMC-MAC is compared to that of RBAR.

The simulation results in Table 5.2 tell that OMC-MAC achieves better performance when the channel diversity F is 2 or 3. We note that the differences between the analysis and the simulation results are due to the control overhead.

Table 5.2. OMC-MAC throughput gain vs. number of channels (L=5)

M (F)	2 (4)	3 (3)	4 (2)	5 (1)
Analysis	$2\sum_{i=1}^{4}\frac{1}{i} = 4.2$	$3\sum_{i=1}^{3}\frac{1}{i} = 5.5$	6	5
Simulated	3.3	4.5	5.1	4.2

5.6. Using Particle Filtering to Estimate the Number of Competing Stations

5.6.1. Sampling Importance Resampling Particle Filter

In this section, we explore the adaptive contention window mechanism for OMC-MAC, which requires online estimation of the number of competing stations. We propose to use the particle filter for this purpose. Roughly speaking, particle filtering is a Bayesian technique used to build the posterior probability density function (PDF) of the state. To illustrate the basic ideal of particle filtering, consider the following dynamic system modeled in a state-space form:

$$State\ evolution: \quad x_t = f_t(x_{t-1}, u_{t-1}), \tag{5.10}$$

$$Measurement: \quad z_t = h_t(x_t, v_t), \tag{5.11}$$

where x_t is the state at time t, z_t is the measurement, u_t is the state noise and v_t is the measurement noise. Suppose that a function of x_t, denoted by $g(x_t)$, is to be estimated. The optimal estimation based on the current observation is $\mathbb{E}[g(x_t)|z_t] = \int g(x_t)p(x_t|z_t)dx_t$, where $p(x_t|z_t)$ is the posterior PDF. In most cases, however, this optimal solution is intractable due to the complexity of the dynamic system. Instead, via *importance sampling*, one can use samples (refereed as "particles") and associated weights to construct the required posterior PDF. Observing that the state-space model exhibits Markovian structures, we can recursively update the state estimation using new measurements, and this can be done using the Sequential Importance Sampling (SIS) method.

Loosely, the particles can be generated from some other distribution $q(x_t|x_{t-1}, z_t)$, which would be easier for sampling, compared to the target $p(x_t|z_t)$. Usually, $q(x_t|x_{t-1}, z_t)$ is called the importance density. The corresponding weights are obtained by

$$w_t^i = w_{t-1}^i \frac{p(z_t|x_t)p(x_t|x_{t-1})}{q(x_t|x_{t-1}, z_t)} \tag{5.12}$$

and the posterior PDF $p(x_t|z_t)$ can be approximated as

$$p(x_t|z_t) \approx \sum_{i=1}^{N} w_t^i \delta(x_t - x_t^i). \tag{5.13}$$

When the number of particles N grows, this approximation approaches the true density function $p(x_t|z_t)$ [7]. Typically, we use $p(x_t|x_{t-1})$ for $q(x_t|x_{t-1}, z_t)$ to simplify the implementation, and (5.12) becomes

$$w_t^i = w_{t-1}^i p(z_t|x_t). \tag{5.14}$$

The SIS filter is known to have the degeneracy problem, which states that many of the particles would have negligible weights after some iterations [7]. The degeneracy problems can make the Monte Carlo procedure inefficient, simply because a large amount of the computation time has been devoted to update particles which has no contribution to the approximation of posterior density. Fundamentally, we can increase the number of particles to reduce this effect. An alternative approach is to use resampling, which requires generating N samples from the approximate density function $p(x_t|z_t)$ to eliminate particles that have small weights. After the resampling, all the particles have the same weights as $1/N$. The above procedures is named as the Sampling Importance Resampling (SIR) filter and can be summarized in Alg. 2 (please see [7] for details):

Algorithm 2 SIR particle filter

Procedure:
 1: **for** time k **do**
 2:   **for** $i = 1 : N$ **do**
 3:     Draw samples from the state space model, $x_t^i \sim p(x_t|x_{t-1})$;
 4:     Calculate the weights w_k^i according to the measurement model, i.e., $w_k^i = p(z_t|x_t^i)$
 5:   **end for**
 6:   Normalize $w_k^i = w_k^i / \sum_i w_k^i$;
 7:   Resample $\{x_t^i\}$.
 8: **end for**

5.6.2. State-Space Model for Estimating the Competing Stations

Next, we derive the state-space model based on [20], which is a key step of using particle filtering to estimate the number of competing stations.

From [18], we know that the probability that an intended transmission encounters a collision is given by

$$p = 1 - (1 - \tau)^{n-1}, \qquad (5.15)$$

where τ is given by (5.3).

It follows that the number of competing stations and the collision probability obey the following relationship:

$$p = 1 - (1 - \frac{2}{nK+1})^{n-1}, \tag{5.16}$$

where $W = nK$. Since p can be observed by counting the number of busy slot in a period (see [20] for details), (5.16) gives the measurement model:

$$p_t = 1 - (1 - \frac{2}{n_t K + 1})^{n_t - 1} + v_t, \tag{5.17}$$

where v_t is a Gaussian random variable with zero mean and the variance by [20]

$$\text{var}(v_t) = \frac{(1 - (1 - \frac{2}{n_t K + 1})^{n_t - 1})((1 - \frac{2}{n_t K + 1})^{n_t - 1})}{S}. \tag{5.18}$$

S is the size of the observation window.

The state evolution equation is straightforward and can be given as [20]

$$n_t = n_{t-1} + u_{t-1}, \tag{5.19}$$

where u_t represents the variation of the competing stations. Here we assume that it is a Gaussian random variable with zero mean and variance σ^2.

In summary, using this state-space model, we can use the particle filter in Alg. 2 to estimate the number of competing stations online by observing the collision probability.

5.6.3. Numerical Results

In this section, we evaluate the above estimation algorithm, compared with Kalman filtering. In order to obtain a quantitative comparison, we follow [7] and use the Root Mean Square Error (RMSE) as the performance metric. Each simulation setting has been run for ten times and 200 particles are used in the SIR filter. The comparison of the two filters is summarized in Table 5.3. It can be seen that the SIR particle filter yields better performance than the EKF filter. We can also find that when σ^2 is large, which implies when there is a frequent change of the number of competing stations, the estimations are less accurate. Figure 5.8 shows the relationship of the true states with the posterior mean estimates from the SIR filter. We can see that the simulation results matches the true states very well.

Table 5.3. RMSE of state estimation (n=10): mean (variance)

Algorithms	$\sigma^2 = 0.1$	$\sigma^2 = 0.01$	$\sigma^2 = 0.001$
EKF	6.93 (3.95)	1.52 (0.17)	0.20 (0.011)
SIR	2.64 (1.63)	0.46 (0.04)	0.13 (0.007)

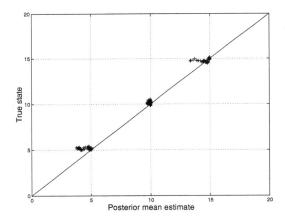

Figure 5.8. True states vs. posterior mean estimates ($\sigma^2 = 0.01$).

5.6.4. Simulation Experiments

In this section, we integrate the above estimation algorithm into OMC-MAC. The simulation settings are the same as those in the second setting of Section 5.5.1.4. Hence, we can learn from the simulation results the performance gain due to the use of the adaptive contention window algorithm compared with the standard BEB algorithm. Fig 5.9 shows the aggregate throughput gain, which ranges from 10% to 13% with an average of approximately 12%, indicating that the adaptation algorithm is effective in improving the throughput performance of OMC-MAC.

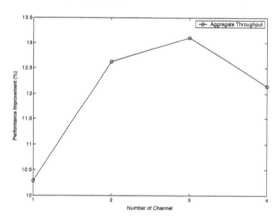

Figure 5.9. Aggregate throughput gain vs. number of channels

5.7. Conclusion

In this chapter, we proposed an opportunistic multi-channel medium access control protocol (OMC-MAC). The OMC-MAC protocol takes into account the channel condition in MAC layer design. More specifically, it exploits the channel variations among multiple channels and selects the best channel for data communications, resulting in multi-channel diversity. It is in this sense we call it an "opportunistic multi-channel MAC protocol". A key feature of OMC-MAC is that the RTS/CTS/RES handshake is used not only for reserving the communication floor, but also for channel allocation and rate adaptation purposes. Performance analysis shows that OMC-MAC can obtain significant multi-channel diversity gain. Furthermore, we proposed to use a sequential Monte Carlo method, namely, the Sampling Importance Resampling particle filter to estimate the number of competing stations, and adjust the contention window size accordingly. Using extensive simulations in GlomoSim, we showed that the proposed OMC-MAC protocol can achieve efficient channel utilization.

Part II

Utility-based Random Access and

Flow Control

CHAPTER 6

A TWO-PHASE UTILITY MAXIMIZATION APPROACH FOR WIRELESS

MEDIUM ACCESS CONTROL

6.1. Introduction

6.1.1. A Network Utility Maximization Framework

The past decade has witnessed an increasing intensity of interest in the utility maximization approach for Internet congestion control, and, more generally, for network resource allocation. Roughly speaking, network control is treated as a distributed solution
to maximizing the utility functions of the users or the network, subject to resource constraints such as capacity constraints or power/energy constraints. The utility functions
serve as a measure of efficiency or fairness, or model user satisfaction or traffic elasticity.
More recently, this utility approach has been substantially generalized from an analytic
tool of reverse-engineering TCP congestion control to a network utility maximization
(NUM) framework that enables a better understanding of interactions across protocol
layers and network architecture choices. This framework of "layering as optimization
decomposition" offers a rigorous integration of the various protocol layers into a single co-
herent theory, by viewing them as carrying out an asynchronous distributed computation
over the network to implicitly solve a global NUM problem. Different layers iterate on
different subsets of the decision variables at different timescales using local information
to achieve individual optimality.

[1] ©[2009] IEEE. Reprinted, with permission, from Dong Zheng and Junshan Zhang, "A Two-Phase
Utility Maximization Framework for Wireless Medium Access Control", IEEE Trans. on Wireless Communications, Vol. 6, Issue 12, Dec. 2007 Page(s):4299-4207.

6.1.2. Challenges in the MAC Design for Wireless Networks

In this chapter, we will take a utility maximization approach for studying medium access control in wireless ad hoc networks [104].

Needless to say, the optimal MAC design in multi-hop wireless networks is challenging, partially due to the time-varying nature of the PHY-layer communication channels and the network topology. Indeed, the unique characteristics of multi-hop networks pose the following two major challenges for MAC design under QoS constraints.

1) **Topology-dependency:** The radio channel in multi-hop wireless networks is spatially shared by many links. Due to the broadcast nature of wireless communications, two active links within a certain distance interfere with each other. Accordingly, only one active transmission is allowed within a certain neighborhood (often modeled as maximal cliques; see [69]). Furthermore, a link can belong to multiple cliques depending on the network topology. As a result, different links may experience different contention levels, making "fair" channel access more challenging.

2) **Channel-dependency:** In many practical wireless systems, there exists a significant amount of multipath fading, and fading is regarded as an intrinsic feature of wireless channels. Since flows experience different channel conditions at different time instances, there is no equivalent notion of well-defined fixed "link bandwidth" in wireless networks as in wireline networks.

One key observation is that in multi-hop wireless networks, the channel variation due to multi-path fading often occurs on a time scale of mili-seconds, whereas the network topology changes relatively slowly since the topology variation is due to mobility, typically on a time-scale of minutes. In light of different time scales of these two variations, it is plausible to "decompose" the MAC design into two phases to address them separately. Thus motivated, we take a two-phase utility approach to study fair MAC design towards QoS provisioning for multi-hop wireless networks; and an ultimate goal here is to develop a utility maximization framework for fair MAC design.

Simply put, in this utility approach, every link attempts to maximize its own utility that is a function of its throughput (we use the $(\vec{w}, \kappa)$-fair utility functions to be defined in Section 6.2), and the optimization procedure takes two phases: A global phase arbitrates fair channel access across the links by adapting their transmission probability, aiming to address the topology dependence and achieve long-term fairness; and a local phase is designed to take into account local channel conditions to achieve short-term fairness, and to determine the transmission duration in each transmission round. Clearly, this scheme has a strong flavor of *cross-layer design*. In what follows, we also call the global phase *the fair channel access phase*, and the local phase *the adaptive data transmission phase*.

One unique contribution of this study is that for the fair channel access phase we use *stochastic approximation* to facilitate the MAC design, based on the fact that the MAC throughput depends on the realizations of channel contention in random access networks. To our best knowledge, there has been little work on using stochastic approximation for MAC design in multi-hop networks, and this paper is aimed at filling this void.

The main contributions of this two-phase utility approach can be summarized as follows.

1) In the fair channel access phase, global utility optimization is carried out to maximize $(\vec{w}, \kappa)$-fair utility functions [68] through distributed adaptation of the transmission probabilities to achieve long-term fairness. We show that the adaptation of the persis-

tence probability can be well modelled by a stochastic approximation algorithm [12]. In particular, we use the weighted proportional fair (WPF) MAC in single-hop networks as an example to illustrate the basic ideas of the stochastic approximation approach. We then generalize the study to establish the stability and fairness of the $(\vec{w}, \kappa)$-fair MAC in multi-hop networks by using LaSalle's Invariance Theorem [66]. We should note that in multi-hop networks some links belong to multiple cliques and the adaptation of persistence probability is coupled across the cliques. Our findings reveal that under the large network assumption, there exists a single equilibrium point for the proposed $(\vec{w}, \kappa)$-fair MAC when $\kappa > 1$.

2) In the adaptive data transmission phase, local utility optimization is carried out to determine the transmission duration while maintaining short-term fairness. To this end, first the transmission rate is chosen based on the channel condition obtained during the RTS/CTS handshake. Then, the transmission duration is determined accordingly by solving the local optimization. The underlying rationale is that the channel utilization can be enhanced by adapting the transmission duration to the channel conditions while taking into account the short-term fairness requirements.

We note that fair access control in wireless networks has received much attention in the past few years. Particularly, fair scheduling algorithms are proposed to achieve weighted fairness among traffic flows sharing the same wireless channel [72, 65], and these algorithms are designed for centralized systems. In contrast, contention-based MAC protocols in [73, 17] attempt to assign equal share of bandwidth to different users with an inherent assumption of equal weights. Recently, backoff-based schemes are presented in [11, 91] to address weighted fair bandwidth allocation. Following the seminal work [52],

[69] presents a novel framework of translating any pre-specified fairness model into a corresponding persistence/backoff-based contention resolution scheme. We will show in Section 6.6 that the proportional fairness scheme in [69] in fact achieves the minimum delay fairness – a special case of our general framework in this paper. Based on the contention graph formulation in [69], [38] develops fair bandwidth sharing algorithms using game theory frameworks. We note that [56, 94] have used utility maximization to study the random access protocol based on deterministic convex programming. Another related work [43] takes a stochastic approximation approach to study distributed multi-access communications in single-hop networks. In contrast, in this paper, we take into account time-varying channel conditions, and study joint MAC and PHY-layer design for multi-hop wireless networks to achieve efficiency and fairness.

The rest of the chapter is organized as follows. Section 6.2, contains the system model. In Section 6.3, we present the problem formulation and use the WPF case to illustrate the basic ideas for MAC design. We examine in details the global phase and the local phase of the distributed MAC design in Sections 6.4 and 6.5. Simulation results are provided in Section 6.6. Finally, we conclude this chapter in Section 6.7.

6.2. System Model

6.2.1. Fairness Notion

In the utility maximization framework [52, 85], different fairness notions correspond to different utility functions. In this paper, we use $(\vec{w}, \kappa)$ fairness with the following utility functions [68]:

$$U_\kappa(\theta_i) = \begin{cases} w_i \log \theta_i, & \text{if } \kappa = 1; \\ w_i(1 - \kappa)^{-1}\theta_i^{1-\kappa}, & \kappa \neq 1. \end{cases} \qquad (6.1)$$

It is clear that these utility functions are *continuously differentiable, increasing, and strictly concave.*

Note that the $(\vec{w}, \kappa)$ fairness boils down to the weighted proportional fairness if $\kappa = 1$, to the minimum delay fairness when $\kappa = 2$, and to the max-min fairness as $\kappa \to \infty$ [85].

6.2.2. Access Control via Utility Optimization

We consider a multi-hop wireless network with a set $\mathcal{S}$ of nodes and a set $\mathcal{L}$ of links, and our focus here is on the MAC design for fair channel access control. For simplicity, we assume that each link is associated with only one transmitter node and one receiver node. We note that our study here can be extended to the case where a node communicates with multiple nodes.

To capture the constraints imposed by the topology dependency in multi-hop networks, *contention graphs* are often used to represent the contention relationship among the links [69]. Specifically, in a contention graph, each vertex stands for an active link, and two vertices are connected by an edge if they interfere with each other. It is understood that only one link can be active in each clique. Then, the cliques reveal the resource constraints of the links and can be derived based on the contention graph [69].

For each link i, let θ_i denote its throughput, x_i the persistence probability of the transmitter node of link i and $U_i(\theta_i)$ the corresponding utility. Let $\mathcal{C}$ be the set of cliques.

The utility optimization problem can then be put as follows:

$$P_1 : \max_{\{\theta_i, i \in \mathcal{L}\}} \quad \sum_i U_i(\theta_i)$$

$$\text{subject to}: \quad \sum_{l:i \in c} x_i \le 1, \forall c \in \mathcal{C} \tag{6.2}$$

$$0 \le x_i \le 1, i \in \mathcal{L},$$

where the rational behind the constraints can be outlined as follows: on average there should be at most one active link in any given clique at any time. We note that in most existing studies on multi-hop networks, the utility functions are defined for end-to-end flows. In contrast, in this study, the utility functions are defined on links, and we investigate fair channel access across the links. It is clear that achieving link-level fairness is also of great interest, and has received much attention (see, e.g., [69, 56] and the references therein).

Observe that from (6.2) the network utilities are functions of the throughput whereas the constraints are in terms of the persistence probabilities. Clearly, a key step is to characterize the dependence of $\{\theta_i\}$ on $\{x_i\}$, which is the subject of the next section.

6.3. Problem Formulation

Consider a homogeneous network where the wireless links have the same channel statistics[1]. We assume that the transmission rates $\{r_{n,i}, \ i \in \mathcal{L}, \ n = 1, 2 \cdots\}$ of the wireless links during round n are independent and identically distributed with mean R. Let U_i and V_i denote the utility functions of link i for the global phase and the local phase, for $i \in \mathcal{L}$ (note that V_i can be different from U_i).

---

[1]We note that the homogeneous case is a basic setting to start with, and the insights obtained here serve as a basis for studying heterogeneous networks.

124

6.3.1. Achieving Long-Term Fairness via Adaptive Persistence Probability

Since the network topology changes relatively slowly, we propose to use the persistence transmission mechanism to address the topology dependence and achieve long-term fairness. More specifically, in the persistence mechanism, every node maintains a persistence probability with which it contends for the channel. In principle, the persistence probability should be adjusted to match the contention level and the fairness requirements.

Assume for now that the transmission duration is fixed at T, and R is normalized to 1. Then, based on the standard results on the persistence algorithm [14, 50], the throughput for link i can be well approximated by $x_i \prod_{j \in I(i)} (1 - x_j)$, where $I(i)$ consists of the links interfering with the transmission of link i.

Clearly, the optimal x_i's are then the solutions to the following optimization problem:

$$P_G: \quad \max_{\{x_i, i \in \mathcal{L}\}} \quad \sum_i U_i(x_i \prod_{j \in I(i)} (1 - x_j))$$

$$\text{subject to}: \quad \sum_{i:i \in c} x_i \leq 1, \forall c \in \mathcal{C} \tag{6.3}$$

$$0 \leq x_i \leq 1, i \in \mathcal{L}.$$

In general, the above optimization problem is non-convex because the objective function involves the product term $x_i \prod_{j \in I(i)} (1 - x_j)$, even though the utility functions $\{U_i\}$ are strictly concave and the constraints are linear.

6.3.2. Achieving Short-Term Fairness via Adaptive Transmission Duration

Once a link (say link i) successfully reserves the channel, the instant channel condition can be estimated using the pilot symbols in the RTS/CTS packets, and the transmission rate can be selected accordingly using adaptive modulation (see [44] for details). The next key step is to determine the transmission duration based on the transmission rate and the short-term fairness requirement. To this end, we model the data transmission in each round as a local optimization problem. Specifically, the transmission duration for the $(n+1)$th transmission round can be obtained by solving the following local problem P_L:

$$P_L : \max_{t_{n+1,i}} \quad \sum_j V_j(\theta_{n+1,j})$$
$$\text{subject to} : \quad 0 \leq t_{n+1,i} \leq t_c, \tag{6.4}$$
$$j \in \{1, 2, \ldots, m\}.$$

where θ_n and T_n evolve as follows:

$$\theta_{n+1,i} = \frac{\theta_{n,i} T_n + t_{n+1,i} r_{n+1,i}}{T_n + t_{n+1,i}},$$
$$T_{n+1} = T_n + t_{n+1,i},$$

and t_c denotes the channel coherence time (within which the channel condition remains more or less unchanged).

6.3.3. Example 1: Weighted Proportional Fair MAC For Single-hop Networks

To illustrate the basic ideas, we now present the two-phase algorithm for the weighted proportional fairness (WPF) MAC, i.e., $U_i(\cdot) = V_i(\cdot) = w_i \log(\cdot)$. For simplicity, we consider a single-hop network with m links (extensions to general $(\vec{w}, \kappa)$-fair MAC in multi-hop networks will be presented in Sections 6.4.3 and 6.5).

We have the following observations for the two-phase utility optimization problem in the WPF case.

1) The global optimization problem turns out to be a constrained convex program. To derive distributed algorithms, we consider the following relaxation of (6.3) using the

penalty method (cf. [85, p. 29]):

$$P_G: \quad \max_{\{0 \le x_i \le 1\}} \sum_i \alpha w_i \log(x_i \prod_{j \ne i}(1 - x_j)) - \beta \int_0^{\sum_j x_j} \lambda(x)dx, \tag{6.5}$$

where $\lambda(x)$ is the penalty function and can be regarded as the "price" for sending data in the clique, and α and β are positive constants. As is standard, we assume that $\lambda(x)$ is a non-negative, non-decreasing, continuous function. A good penalty function for (6.5), in the sense that the solution to (6.5) approximates that to (6.3) well, should have the property that $\lambda(x)$ is very small for $x \le 1$ and becomes large when $x > 1$. We note that the ratio β/α is also intimately related to the approximation.

Appealing to the gradient method, we obtain the following adaptive persistence algorithm for solving (6.5):

$$\dot{x}_i = \alpha w_i - \alpha x_i \sum_j w_j - \beta x_i(1 - x_i)\lambda(\sum_i x_i)), \tag{6.6}$$

for $i \in 1, \ldots, m$.

2) After some algebra, the solution to the local optimization problem (6.4) can be shown to be

$$t_{n+1,i} = \min\left\{ \left[\left(\frac{w_i}{\sum_{j \ne i} w_j} - \frac{\sum_i w_i}{\sum_{j \ne i} w_j} \frac{\theta_{n,i}}{r_{n+1,i}} \right) T_n \right]^+, t_c \right\}.$$

6.4. Utility Maximization in the Global Phase

6.4.1. A Stochastic Approximation Approach for the Global Optimization Phase

To implement (6.6) in random access networks, we seek to understand the following issues:

1. What price function $\lambda(\cdot)$ should be chosen for distributed MAC?

2. How could the price information $\lambda(\cdot)$ be generated and fed back to the links so that the MAC protocol can be executed distributively?

3. Under what condition can the stability and fairness be guaranteed?

Observe that in the wireless medium access control, the feedback to a node is determined by the events it can observe, and the outcome can be success, failure or idle depending on the acknowledgment and channel activity. In particular, two important observations are in order: 1) the feedback information is determined by the sample realization of channel contention; and 2) the overall contention outcomes in a long run reveal the contention level in the clique. That is, a higher frequency of collisions implies that the clique is more likely to be saturated and the links should decrease their transmission probabilities to reduce collisions; in contrast, a lower frequency of collisions indicates that the clique may be under-utilized and the links should increase the transmission probabilities. In summary, the price function, $\lambda(x)$, is determined by the random access mechanism and can be inherently obtained from the collision outcomes (we will see that one price function in the global phase for WPF MAC is simply $\lambda(x) = x$). Thus motivated, we propose to use a stochastic approximation method to "implement" the distributed algorithm.

The standard form of stochastic recursive algorithm is given by [12]

$$\vec{x}(t+1) = \vec{x}(t) + \gamma F(\vec{\epsilon}(t), \vec{x}(t)), \tag{6.7}$$

where $\vec{\epsilon}$ is the random inputs and γ is the step size, and is a small positive constant. It has been shown that under some mild conditions and given that γ is sufficiently small, the stochastic algorithm in (6.7) converges in probability to the attractor of the mean ODE:

$$\frac{d\vec{x}(t)}{dt} = E[F(\vec{\epsilon}, \vec{x}(t))|\vec{x}(t)], \tag{6.8}$$

provided that the mean ODE is globally stable [12, 55].

To facilitate the stability analysis, we impose the following large network assumption.

[**Condition 1**]: *We assume that in a large network, $x_i \ll \sum_{j \in c} x_j$ and $x_i \to 0$ as $|c| \to \infty$, for all $c \in \mathcal{C}$ and $i \in c$, where $|c|$ is the cardinality of clique c.*

Intuitively speaking, Condition 1 requires that in a network with many links, no single link would dominate the channel usage. We note that Condition 1 is applicable to some practical scenarios of interest, e.g., sensor networks where a large number of sensor nodes are deployed.

6.4.2. Example 1 (cont'd): Algorithm Description and Stability Analysis for WPF-MAC

We continue to use the WPF case to illustrate the implementation of the adaptive persistence algorithm using stochastic approximation. More specifically, combining (6.6), (6.7) and (6.8), we propose the following "sample path" adaptive algorithm for the WPF case:

$$x_{n+1,i} = x_{n,i} + \alpha w_i - \alpha \sum_j w_j \cdot \mathbf{I}(\mathcal{E}_i) - \beta(1 - x_{n,i}) \cdot \mathbf{I}(\mathcal{F}_i) \cdot \mathbf{I}(\mathcal{E}_i), \qquad (6.9)$$

where $\mathcal{E}_i$ denotes the event that node i contends for the channel; and $\mathcal{F}_i$ corresponds to the event that there is a collision with node i's transmission. Note that α and β are positive step sizes, which are chosen small enough to incorporate the effect of γ.

Next, we elaborate further on the algorithms corresponding to (6.9). Specifically, in each transmission round, node i contends for the channel with probability $x_{n,i}$ and updates its probability according to the contention outcome by the following policy (see Fig. 6.1)
:

   1. Increase $x_{n,i}$ by αw_i, regardless it contends or not;

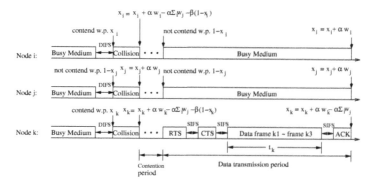

Figure 6.1. A sketch of one transmission round of WPF-MAC

2. If it contends and succeeds, decrease $x_{n,i}$ by $\alpha \sum_j w_j$;

3. If it contends and fails, decrease $x_{n,i}$ by $\alpha \sum_j w_j + \beta(1 - x_{n,i})$.

We now examine the convergence of this algorithm and analyze the fairness therein. It can be shown that the mean ODE for (6.9) (up to a constant γ) is given by

$$\dot{x}_i = \alpha w_i - \alpha \sum_j w_j E[\mathbf{I}(\mathcal{E}_i)|\vec{x}] - \beta E[(1 - x_i)\mathbf{I}(\mathcal{F}_i)\mathbf{I}(\mathcal{E}_i)|\vec{x}]. \tag{6.10}$$

Under Condition 1, we have that

$$E[\mathbf{I}(\mathcal{F}_i)|\vec{x}] \approx 1 - \exp(-\sum_j x_j). \tag{6.11}$$

Based on Taylor's expansion, we "neglect" the terms of high order (> 2) and approximate the mean ODE (6.10) as

$$\dot{x}_i = \alpha w_i - \alpha x_i \sum_j w_j - \beta x_i(1 - x_i) \sum_j x_j, \tag{6.12}$$

which is the same as (6.6) with $\lambda(\sum_i x_i)$ being $\sum_i x_i$. That is, under Condition 1, the penalty function has a simple form as $\lambda(x) = x$ for the adaptive persistence algorithm.

Appealing to Theorem 7.2.1 in [55, p. 218], we have the following proposition regarding the fairness and stability of the stochastic algorithm (6.9).

Proposition 6.4.1. Under Condition 1, the stochastic algorithm in (6.9) converges in probability to the unique equilibrium point of (6.12) that maximizes the following Lyapunov function:

$$L(X) = \sum_i \alpha w_i \log \left(x_i \prod_{j \neq i} (1 - x_j) \right) - \beta \int_0^{\sum_j x_j} x \, dx.$$

The proof of Proposition 6.4.1 follows the standard procedure of stochastic approximation [55, 12], and a key step here is to establish the convergence of the mean ODE (6.12), which can be verified using Lyapunov's Direct Method (cf. [85, p. 46]).

We note that, in general, it is difficult to obtain the close form of the equilibrium point of (6.12). Nevertheless, we have the following proposition on the successful transmission probability at the equilibrium point. For convenience, let $X_0 = [x_{10}, \ldots, x_{m0}]^T$ denote the equilibrium point and define $\mu_m \triangleq \prod_{j=1}^m (1 - x_{j0})$.

Proposition 6.4.2. *Under Condition 1, the successful transmission probability P_s is lower-bounded by p, where $P_s \triangleq \sum_i x_{i0} \prod_{j \neq i} (1 - x_{j0})$ and $p \triangleq \frac{\sum_i w_i}{\sum_i w_i + \beta/\alpha} \mu_m$.*

A key observation from Proposition 6.4.2 is that the lower-bound p on P_s is a decreasing function of the ratio β/α, i.e., the smaller β/α is, the larger p becomes. However, recall that the larger β/α is, the closer X_0 is to the solution of Problem P_G, indicating that there is a trade-off between throughput and fairness.

6.4.3. Generalizations to Multi-hop Networks with $(\vec{w}, \kappa)$ Fairness: Algorithm Description and Stability Analysis

In previous sections, we have illustrated the MAC design for the WPF case in a single-hop network. Built on the above studies, the $(\vec{w}, \kappa)$ fairness can be achieved by applying the $(\vec{w}, \kappa)$-fair utility functions (6.1) to (6.3) and (6.4) respectively. We should note that some links belong to multiple cliques and the adaptation of persistence probability is coupled across the cliques. We elaborate further on this in the following. To facilitate stability analysis, we assume that $\kappa > 1$ in what follows.

We first outline a few key challenges in applying the general $(\vec{w}, \kappa)$-fair model. Firstly, in the global phase, the overall objective function may not be concave due to the product term involving x_i's. And more importantly, since the objective function is not separable, the globally optimal points can be extremely difficult to achieve [13]. As noted in [85], an exact solution to the original problem may not be necessary because the network models are often approximations of real networks. Therefore, we seek relaxations of (6.3) so that approximate solutions to (6.3) can be found via distributed algorithms. We first have the following lemma.

Lemma 6.4.1. For all $c \in \mathcal{C}$, if $\max_{j \in c} x_j \to 0$ and $\sum_{j \in c} x_j \to \mu_c$ as $|c| \to \infty$, then $\prod_{j \in c}(1 - x_j) \to e^{-\mu_c}$ as $|c| \to \infty$.

Based on Lemma 6.4.1 and (6.1), we have that, when $|I(i)|$ is sufficiently large,

$$U_{\kappa,i}\left(x_i \prod_{j \in I(i)} (1 - x_j)\right) - U_{\kappa,i}\left(x_i e^{-\mu_c}\right)$$

$$= w_i(1 - \kappa)^{-1} x_i^{1-\kappa} \left[\left(\prod_{j \in I(i)} (1 - x_j)\right)^{1-\kappa} - \left(e^{-\mu_c}\right)^{1-\kappa}\right]$$

$$\overset{(a)}{\leq} w_i x_i^{1-\kappa} \left(\prod_{j \in I(i)} (1 - x_j) - e^{-\mu_c}\right) e^{-\mu_c \kappa}$$

$$\to 0,$$

where (a) comes from the concavity of the utility function $(1 - \kappa)^{-1} x^{1-\kappa}$. With the above approximation, problem (6.3) can be converted into the following convex programming problem:

$$P_G: \max_{\{x_i\}} \quad \sum_i w_i'(1 - \kappa)^{-1} x_i^{1-\kappa}$$

$$\text{subject to}: \quad \sum_{i: i \in c} x_i \leq 1, \forall c \in \mathcal{C}, \tag{6.13}$$

$$0 \leq x_i \leq 1, i \in \mathcal{L},$$

with $w_i' = w_i e^{\mu_c(\kappa-1)}$.

Along the same line as in the WPF case, we propose the following stochastic approximation algorithm for the $(\vec{w}, \kappa)$-fair case:

$$x_{n+1,i} = x_{n,i} + \alpha w_i' - \beta x_{n,i}^{\kappa-1} \cdot \mathbf{I}(\mathcal{F}_i) \cdot \mathbf{I}(\mathcal{E}_i). \tag{6.14}$$

Again, based on Taylor's expansion, we "neglect" the terms of high order (> 2) and approximate the mean ODE of (6.14) as follows:

$$\dot{x}_i = \alpha w_i' - \beta x_i^\kappa \sum_{j \in I(i)} x_j. \tag{6.15}$$

Note that the price information in (6.15) is generated by the sum of the probabilities of all links in $I(i)$, which indeed captures the topology-dependent contention in multi-hop wireless networks.

Let $\alpha \leq \beta$ and $w'_i \leq 1$. Based on (6.15), we define a corresponding Lyapunov function as follows:

$$L(X) = \sum_i \alpha w'_i (1 - \kappa)^{-1} x_i^{1-\kappa} - \frac{1}{2} \sum_{i,j} a_{ij} x_i x_j, \qquad (6.16)$$

where $a_{ij} = \beta, i \neq j$ if i and j are in the same clique, otherwise $a_{ij} = 0$. Intuitively, the coefficients $\{a_{ij}\}$ capture the topology dependence. Our main result on the stability and fairness of the $(\vec{w}, \kappa)$-fair MAC scheme is stated as follows:

Proposition 6.4.3. Under Condition 1, the stochastic algorithm in (6.14) converges in probability to the single equilibrium point of (6.15), which is the solution to $\dot{L}(X) = 0$ with $L(\cdot)$ given by (6.16), provided that $\kappa > 1$.

Proof: We prove Proposition 6.4.3 in two steps. *Step I*: In light of the structure of the Lyapunov function in (6.16), we use LaSalle's Invariance Theorem [66, p. 90] to establish the convergence of the system in (6.15), i.e., we show that (6.15) converges to the largest invariant set N in $E \triangleq \{x : \dot{L}(x) = 0, x \in [0, 1]^n\}$; *Step II*: we prove that there is only one single point in N.

Step I: It is clear that $L(X)$ defined in (6.16) is a upper-bounded continuously differentiable function in the compact set $[0, 1]^n$, and furthermore, $\dot{L} \geq 0$ because of (6.15). Based on LaSalle's Invariance Theorem [66, p. 90], every solution of (6.15) must converge to the largest invariant set N in E.

Step II: From (6.15), it is not difficult to see that E is non-empty because $\alpha \leq \beta$ and $w'_i \leq 1, \forall i \in \mathcal{L}$. Next, we show E has only one element. We prove this by contradiction.

Assume there are m links in $\mathcal{L}$. Suppose there are more than one equilibrium points. Let $X = [x_1, \ldots, x_m]^T$ and $X' = [x'_1, \ldots, x'_m]^T$ be two of them. We note that the value of

134

$x_i, \forall i \in \{1, \ldots, m\}$ cannot be 0 at the equilibrium point because of the positive drift αw_i in (6.15).

For convenience, define

$$\rho \triangleq \max\{\frac{x_i}{x'_i}, \forall i \in \{1, \ldots, m\}\}.$$

We next show that $\rho > 1$. If $\rho \leq 1$, then $x_i \leq x'_i, \forall i \in \{1, \ldots, m\}$. Since $X \neq X'$, there must exist a link j that $x_j < x'_j$. Because X and X' are equilibrium points, from (6.15), we have that

$$\alpha w'_j = \beta x_j^\kappa \sum_{i \in I(j)} x_i < \beta x'^\kappa_j \sum_{i \in I(j)} x'_i = \alpha w'_j.$$

leading to a contradiction.

Without loss of generality, we assume that $\frac{x_1}{x'_1} \leq \frac{x_2}{x'_2} \leq \cdots \leq \frac{x_m}{x'_m}$ and $x_m/x'_m = \rho > 1$. From (6.15), we have that $\alpha w'_m = \beta x_m^\kappa \sum_{j \in I(m)} x_j = \beta x'^\kappa_m \sum_{j \in I(m)} x'_j$, or equivalently,

$$\frac{x'_{j_1} + x'_{j_2} + \cdots + x'_{j_n}}{x_{j_1} + x_{j_2} + \cdots + x_{j_n}} = \frac{x_m^\kappa}{x'^\kappa_m} = \rho^\kappa, \tag{6.17}$$

where $j_k \in I(m), \forall\, k = 1, 2, \ldots, n$.

Arrange $\{\frac{x'_{j_k}}{x_{j_k}}\}$ in an ascending order. Again, without loss of generality, we assume that

$$\frac{x'_{j_1}}{x_{j_1}} \leq \frac{x'_{j_2}}{x_{j_2}} \leq \cdots \leq \frac{x'_{j_n}}{x_{j_n}}. \tag{6.18}$$

It is easy to show that $\frac{x'_{j_1} + x'_{j_2}}{x_{j_1} + x_{j_2}} \leq \frac{x'_{j_2}}{x_{j_2}}$. Combining this with (6.18), we have that

$$\frac{x'_{j_1} + x'_{j_2} + \cdots + x'_{j_n}}{x_{j_1} + x_{j_2} + \cdots + x_{j_n}} \leq \frac{x'_{j_n}}{x_{j_n}}. \tag{6.19}$$

Combining (6.19) and (6.17) further yields that

$$\frac{x'_{j_n}}{x_{j_n}} \geq \frac{x_m^\kappa}{x'^\kappa_m} = \rho^\kappa. \tag{6.20}$$

Similarly, using (6.15) and the equilibrium condition at j_n, we obtain that

$$\frac{x_m + \sum_{l:l \neq m, l \in I(j_n)} x_l}{x'_m + \sum_{l:l \neq m, l \in I(j_n)} x'_l} = \frac{x'^{\kappa}_{j_n}}{x^{\kappa}_{j_n}}. \tag{6.21}$$

Since $\frac{x_m}{x'_m}$ is the largest, it follows that

$$\frac{x_m}{x'_m} \geq \frac{x'^{\kappa}_{j_n}}{x^{\kappa}_{j_n}}. \tag{6.22}$$

Comparing (6.20) with (6.22), we have that

$$\rho \geq \rho^{\kappa^2} \implies \rho \leq 1,$$

simply because $\kappa > 1$. This contradicts the fact that $\rho > 1$. We conclude that E is a singleton.

Since N is the largest invariant set in E, N has to be a singleton too. Combining the above with standard stochastic approximation arguments, we can conclude the convergence of the $(\vec{w}, \kappa)$-fair MAC in (6.14) to a single equilibrium point. ∎

6.4.4. Numerical Examples

In what follows, we provide numerical examples to corroborate the results on the adaptation of the persistence probability.

Example 2: We first examine the rate of convergence of the proposed algorithm for different α and β. We simulate a WPF case with $m = 10$ links in a single clique. We plot in Fig. 6.2 the difference between the realizations of the persistence probability vector and the equilibrium point, in terms of the sum of square error (SSE). It can be observed that the larger α and β are, the faster the algorithm converges. It should be noted, however, that the fluctuation in the steady state may become more significant as α and β increase.

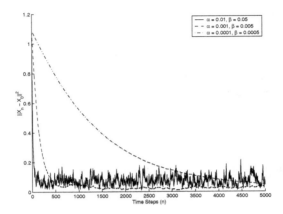

Figure 6.2. Convergence rate of the persistence probability adaptation - the WPF case

That is to say, there exists a trade-off between steady-state performance and fast tracking capability of network conditions.

Example 3: In the second example, we examine the convergence of the $(\vec{w}, \kappa)$-fair MAC in multihop networks. Specifically, we simulate a minimum potential delay fair MAC with 5 links, i.e., $\vec{w} = \{1/5, 1/5, 1/5, 1/5, 1/5\}$ and $\kappa = 2$. The contention matrix A is given as:

$$
A = \begin{bmatrix}
0 & 1 & 1 & 1 & 1 \\
1 & 0 & 0 & 0 & 0 \\
1 & 0 & 0 & 0 & 1 \\
1 & 0 & 0 & 0 & 1 \\
1 & 0 & 1 & 1 & 0
\end{bmatrix},
$$

where it is understood that $a_{i,j} = 1$ if i and j are in the same clique. It can further be shown that the equilibrium point is $X_0 = \{0.1756, 0.4590, 0.3140, 0.3140, 0.2237\}$.

The experiment is repeated for sufficiently many times, and the simulation results corroborate that there is only single equilibrium point for the stochastic algorithm in

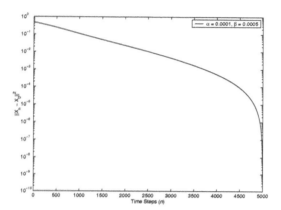

Figure 6.3. Convergence rate of the persistence probability adaptation - the $(\vec{w}, 2)$-fair case

(6.14). To get a more concrete understanding of the convergence, we plot in Fig 6.3 the difference between the persistence probability vector X_n and the equilibrium point X_0, in terms of SSE.

6.5. Utility Maximization in The Local Phase

Next, we turn to the utility maximization of the local phase for the $(\vec{w}, \kappa)$-fair case. After some algebra, the optimal solution to the local optimization problem P_L for the $(\vec{w}, \kappa)$-fair case can be shown as follows:

$$t_{n+1,i} = \min \left\{ \left[\left(\left(\frac{w_i(r_{n+1,i} - \theta_{n,i})}{\sum\limits_{j \in I(i)} w_j \theta_{n,j}^{1-k}} \right)^{\frac{1}{k}} - \theta_{n,i} \right) \frac{T_n}{r_{n+1,i}} \right]^+ , t_c \right\}. \tag{6.23}$$

It can be easily seen from (6.23) that the optimum transmission duration of link i is determined by its channel condition, its throughput, and the throughput of the interfering links. Note that this throughput information can be embedded in the RTC/CTS dialogue so that it can be retrieved by the neighboring links. But for the WPF case, this throughput information is not needed, and a simpler algorithm can be developed for the local phase.

138

Example 1 (cont'd): Local Phase for WPF-MAC:

To ensure short-term fairness, we use a sliding window algorithm with a window size $T^* \gg t_c$, and the transmission duration and the throughput are updated by

$$t_{n+1,i} = \min\left\{ \left[\left(\frac{w_i}{\sum_{j \neq i} w_j} - \frac{\sum_i w_i}{\sum_{j \neq i} w_j} \frac{\theta_{n,i}}{r_{n+1,i}} \right) T^* \right]^+, t_c \right\} \tag{6.24}$$

and

$$\theta_{n+1,j} = \begin{cases} \left(1 - \frac{t_{n+1,i}}{T^* + t_{n+1,i}}\right) \theta_{n+1,i} + \frac{\theta_{n+1,i}}{T^* + t_{n+1,i}} r_{n+1,i} & j = i \\ \left(1 - \frac{t_{n+1,i}}{T^* + t_{n+1,i}}\right) \theta_{n+1,j} & j \neq i. \end{cases}$$

A few important observations are in order.

1) It can be seen from (6.24) that the optimum transmission rate $t_{n+1,i}$ is an increasing function of $r_{n+1,i}$, indicating a larger transmission duration when the channel condition is good and vice versa. We can also observe that $t_{n+1,i}$ is a decreasing function of $\theta_{n,i}$. As a result, the nodes with lower throughput would have a relatively longer transmission duration.

2) Clearly, T^* would impact the fairness and the channel utilization, and should be chosen carefully. We will elaborate on this in Section 6.6.

3) In the implementation, the calculation of the transmission time is performed at the receiver since the transmitter has no knowledge of the channel condition before its transmission, and the CTS packet carries the right duration and the transmission rate. After the transmitter receives the CTS packet, it sends the data packet together with the information of the transmission duration embedded in a special MAC sub-header (see [44]). Therefore, the neighboring nodes of the transmitter need to decode the header only to update their NAVs.

4) It is worth noting that a longer transmission time may require a larger frame size, which may increase the frame error rate. However, by choosing the modulation and the

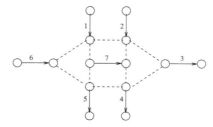

Figure 6.4. A sketch of the multihop network topology

transmission rate carefully, the desired frame error rate can be achieved.

6.6. Simulation Study

The simulation study is organized into two parts. In the first part, we study the WPF-MAC for a multihop network; and in the second part, we examine the performance of the $(\vec{w}, \kappa)$-fair MAC for WLANs. The simulations are performed using GloMoSim [10]. Constant Bit Rate (CBR) traffic is generated from the source nodes. We assume that the channel fading follows the Rayleigh distribution with a coherence time of 20 ms. The transmission rates of data streams can be 2 Mbps, 5.5Mbps, and 11 Mbps, corresponding to the channel conditions. We set $\alpha = 0.001$ and $\beta = 0.005$.

6.6.1. The WPF-MAC for Multihop Networks

We consider a basic multihop network scenario shown in Fig. 6.4. The distances among these links are set such that there are six cliques, i.e., $c_1 = \{1, 2, 3, 7\}$, $c_2 = \{2, 3, 4, 7\}$, $c_3 = \{3, 4, 5, 7\}$, $c_4 = \{4, 5, 6, 7\}$, $c_5 = \{1, 5, 6, 7\}$ and $c_6 = \{1, 2, 6, 7\}$.

We examine the fairness performance of the proposed WPF-MAC under two different settings, one with equal weights for all links, and the other with different weights. The

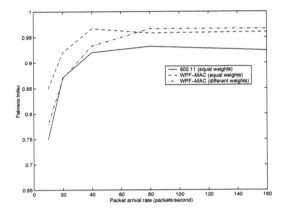

Figure 6.5. The fairness performance of the WPF-MAC

results are plotted in Fig. 6.5 using the weighted proportional fairness index [91]. Note that we use the fairness index as the performance metric, for the sake of comparison with the IEEE 802.11 DCF. Since the IEEE 802.11 DCF does not support the weighted fairness, we only evaluate its fairness performance for the equal-weight setting. It can be seen that the proposed WPF-MAC achieves good fairness performance under both settings, and performs better than the IEEE 802.11 DCF under the equal-weight setting.

6.6.2. The $(\vec{w}, \kappa)$-fair MAC for WLANs

In this section, to study the performance of the $(\vec{w}, \kappa)$-fair MAC, we consider a WLAN setting where there are 5 flows with different transmission weights ($\vec{w} = \{0.32, 0.32, 0.2, 0.12, 0.04\}$).

The fairness performance is examined using the following fairness index adapted from [11]:

$$FI_\kappa = \max_{\{i,j\}} \left\{ \max\left(\frac{\theta_i^\kappa}{w_i}, \frac{\theta_j^\kappa}{w_j}\right) / \min\left(\frac{\theta_i^\kappa}{w_i}, \frac{\theta_j^\kappa}{w_j}\right) \right\}.$$

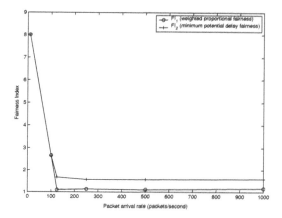

Figure 6.6. The fairness performance of $\{\vec{w}, \kappa\}$-fair MAC ($\{\kappa = 1, 2\}$)

Using the fairness index (FI_κ), we plot in Fig. 6.6 the performance of the $(\vec{w}, \kappa)$-fair MAC when $\{\kappa = 1, 2\}$. From Fig. 6.6, we observe that in the $(\vec{w}, 1)$-fair MAC scheme, the FI_1 is close to 1 when the packet arrival rate is high. We should also note that when the packet arrival rate is small, all the flows achieve almost the same throughput, and this is because of possibly empty queues for higher priority flows. We observe from Fig. 6.6 that the PFCR scheme proposed in [69] in fact minimizes the FI_2 fairness index which achieves the minimum delay fairness [85].

Table 6.1 illustrates the performance gain of the $(\vec{w}, 1)$-fair MAC by using adaptive transmission duration, compared to the no adaptation case. It can be seen that the improvement of the throughput performance is significant for all the links given that T^* is properly chosen.

From Fig. 6.7, it is clear that the sliding window size T^* is important for achieving both long-term and short-term fairness. Recall that $t_{n+1,i}$ is determined by the accumulated throughput $\theta_{n,i}$ and the weights $\{w_i\}$. If T^* is too small, then $t_{n+1,i}$ is given

Table 6.1. Throughput gain of $\{\vec{w}, 1\}$-fair MAC by using adaptive transmission duration

Nodes:	1	2	3	4	5
$T^* = 20MS$	78.9%	92.7%	57.7%	25.2%	8.10%
$T^* = 50MS$	81.7%	80.5%	63.6%	56.0%	47.3%
$T^* = 100MS$	80.8%	83.0%	79.0%	82.4%	85.0%
$T^* = 200MS$	74.3%	76.6%	68.3%	82.0%	83.3%

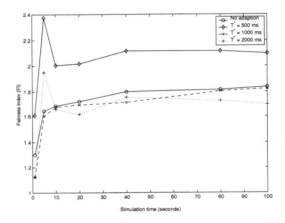

Figure 6.7. $\{\vec{w}, 1\}$-fair MAC fairness performance with respect to the sliding window size (T^*)

by $(w_i / \sum_{j \neq i} w_j) T_n$; as a result, the higher priority flows may occupy the channel longer than needed, resulting in possible unfairness (as illustrated in Fig. 6.7). In contrast, if T^* is too large (e.g., $T^* = 2000$ ms), the short-term fairness may become poorer, because a small positive difference of the first two terms in (6.24) would yield large $t_{n+1,i}$ in this case.

Next, we study the impact of mobility on fairness. Specifically, the *random waypoint* mobility model [10] is used, where the maximum speed is denoted by v_m and the pause-time by t_p. As indicated by the results in Table 6.2, the impact is not significant.

Table 6.2. $\{\vec{w}, 1\}$-fair MAC fairness performance under mobility

	$t_p = 1s$	$t_p = 5s$	$t_p = 10s$
$v_m = 1m/s$	1.6450	1.6269	1.6634
$v_m = 5m/s$	1.6677	1.7178	1.7581
$v_m = 10m/s$	1.6536	1.5327	1.5563
$v_m = 100m/s$	1.5394	1.5642	1.4991

6.7. Conclusion

In this chapter, we proposed a two-phase utility maximization approach to study wireless MAC design towards QoS provisioning. Based on the fact that the topology variation and channel variation occur on different time scales, the utility optimization is "decoupled" into two phases, i.e., the global phase and the local phase. More specifically, in the global phase, the adaptive persistence mechanism is used to achieve the long-term fairness; and in the local phase, the transmission duration is adapted based on the channel condition and the QoS constraints, aiming to achieve the short-term fairness. Under this two-phase utility maximization framework, we presented the detailed algorithm for the weighted proportional fair MAC. In particular, we utilized a stochastic approximation method to design the adaptive persistence mechanism for the global phase, and establish the stability and analyze the fairness therein. Our findings reveal that under the large network approximation, there exists only a single equilibrium point for the proposed $\{\vec{w}, \kappa\}$-fair MAC provided that $\kappa > 1$. We also derived the solution to the local phase under general fairness requirements, and discuss the related implementation issues. We note that the overall optimization for both the global phase and the local phase still remains open. Nonetheless, it is clear that these two algorithms achieve the desired fairness and enhance the channel utilization significantly, as illustrated by simulation results.

CHAPTER 7

THE IMPACT OF STOCHASTIC NOISY FEEDBACK ON DISTRIBUTED

NETWORK UTILITY MAXIMIZATION

7.1. Introduction

In the previous chapter, we have used the network utility maximization (NUM) approach for studying fair MAC design. In this chapter, we study joint optimization of flow control and random access for end-to-end QoS provisioning in wireless ad hoc networks.

7.1.1. Motivation

Distributed solutions play a critical role in making the NUM framework attractive in practical networks where a centralized solution is often infeasible, non-scalable, too costly, or too fragile. The implementation of such distributed algorithms hinges heavily on the communications (message passing) among network elements (e.g., nodes and links), which take place in the form of information exchange and feedback. Unfortunately, in practical systems the feedback is often obtained using error-prone measurement mechanisms and also suffers from other random errors in its transmissions. That is to say, the feedback signals are stochastically noisy in nature. This gives rise to the following important questions:

- Would it still be possible for the distributed NUM solution, in the presence of stochastic noisy feedback, to converge to the optimal point obtained by the corresponding deterministic algorithms; and if yes, under what conditions and how quickly does it converge?

[1]©[2009] IEEE. Reprinted, with permission, from Junshan Zhang, Dong Zheng and Mung Chiang, "The Impact of Stochastic Noisy Feedback on Distributed Network Utility Maximization", IEEE Trans. on Information Theory, Vol. 54, Issue 2, Feb. 2008 Page(s):645-665.

- In the case that it does not converge to the optimum point, would the distributed NUM solution be stable at all?

- Which decomposition method would yield more robust solutions, and what is the corresponding complexity?

A systematic study of the impacts of stochastic noisy feedback on distributed network utility maximization is lacking in the existing literature that almost always assumes ideal feedback. A main objective of this study is to fill this void and to obtain some overriding principles for network resource allocation in the presence of stochastic noisy feedback.

In this chapter, we consider distributed NUM under stochastic noisy feedback in wireline Internet and multi-hop wireless networks. Noisy feedback is ubiquitous in both wireline networks and wireless networks, although its impact in wireless systems is likely to be more significant. Indeed, we observe that the randomness in the noisy feedback signals has its root in a number of error-prone measurement mechanisms inherent in multi-hop networks, as illustrated by two examples we will explore in detail.

As the first example, in the Internet congestion control scheme, the congestion price from the routing nodes needs to be fed back to the source node for rate adaptation. A popular approach for obtaining the price information is the packet marking technique. Simply put, packets are marked with certain probabilities that reflect the congestion level at routing nodes. The overall marking probability depends on the marking events along the routing path that the packets traverse, and is estimated by observing the relative frequency of marked packets during a pre-specified time window at the source node. Clearly, such a feedback signal is probabilistic in nature.

As another important case, consider a wireless network based on random access, where packet transmissions take place over collision channels, and the link feedback information is $(0, 1, e)$. Therefore, the measurement of the flow rate is sample-path-based and depends highly on the specifics of the random access mechanism and the delay can fluctuate significantly. Furthermore, the feedback signal is transmitted over the wireless channel, which is error-prone due to the channel variations of the link quality.

Researchers have recently started to appreciate that there are many decomposition methods that can solve the NUM problem in different but all distributed manners. We first investigate the impact of noise feedback on the distributed NUM algorithms based on the Lagrangian dual method. These algorithms, often regarded gradient or sub-gradient based algorithms, can be roughly classified into the following three categories: primal-dual (P-D) algorithms, primal algorithms and dual algorithms. We shall focus on the impact of noise feedback on the primal-dual algorithm, which is a single time-scale algorithm in the sense that the primal variables and dual variables are updated at the same time. Next, we study distributed two time-scale algorithms based on primal decomposition, and explore the impact of noisy feedback on stochastic stability therein. This study enables us to compare alternative decompositions by the important metric of robustness to noisy feedback.

7.1.2. Overview of results

Assuming that strong duality holds, we study stochastic stability of primal-dual algorithms in the presence of noisy feedback. To this end, we examine the structure of the stochastic gradients (sub-gradients) that are the key elements in distributed NUM algorithms. Since in practical systems, the stochastic gradients can be either biased (e.g., when random exponential marking is used) or unbiased (e.g., when self-normalized additive marking is used), we investigate the stability for both cases: 1) For the unbiased case, we establish, via a combination of the stochastic Lyapunov Stability Theorem and local analysis, that the iterates generated by distributed P-D algorithms converge with probability one to the optimal solution corresponding to the deterministic algorithm, under standard technical conditions. 2) In contrast, when the gradient estimator is biased, we cannot hope to obtain almost sure convergence of the iterates to the optimal points. Nevertheless, we are able to show that the iterates still move along the direction towards the optimal point and converge to a contraction region around the optimal point, provided that the biased terms are asymptotically bounded by a scaled version ($0 \leq \eta < 1$) of the true gradients. Our findings confirm that for a fixed network configuration the distributed NUM algorithms using constant step sizes are less robust to the random perturbation, in the sense that the fluctuation of the iterates remains significant after many iterations; in contrast, the fluctuation using diminishing step sizes vanishes eventually, even under stochastic noisy feedback. However, we caution that in scenarios where time-scale decomposition does not hold well, adaptive step size (not necessarily diminishing) is needed to track the network dynamics since diminishing step sizes may slow down convergence [45, 55].

We also investigate the rate of convergence for the unbiased case, and our results reveal that due to the link constraints, the limit process for the interpolated process corresponding to the normalized iterate sequence, is in general a stationary reflected linear diffusion process, not necessarily the standard Gaussian diffusion process (cf. [52]).

Indeed, the reflection terms incurred by the link constraints help to increase the rate of convergence, and the spread (fluctuation) around the equilibrium point is typically smaller than the unconstrained case. Our result on the rate of convergence can be viewed as a generalization of that in the seminal work by Kelly, Maulloo, and Tan [52]. Analysis of the rate of convergence also serves as a basis for determining time scale separability in network protocol design.

To get a more concrete sense of the above results, we then apply the general theory developed above to investigate stability of cross-layer rate control for joint flow control and MAC design. Specifically, we consider rate control in multi-hop wireless networks based on random access, and take into account the rate constraints imposed by the MAC/PHY layer. Appealing to the general theory above, we examine the convergence behavior of the stochastic algorithm for cross-layer rate control for both unbiased and biased cases. Our numerical examples corroborate the theoretic findings well.

Finally, we generalize the study to explore the stability behavior of distributed two time-scale NUM algorithms based on primal decomposition; In such algorithms, gradient estimators in both faster and slower time scales can be unbiased or biased. When the gradient estimators in both time scales are unbiased, we establish, via the mean ODE method, the convergence of the stochastic two time-scale algorithm under mild conditions. For the case where the gradient estimator at the faster time scale is unbiased and that at the slower time scale is biased, it can be shown that algorithm converges to a contraction region. We should point out that stability is not well defined when the gradient estimator at the faster time scale is biased, because the cumulative effect of the biases can vary significantly. We use numerical examples to illustrate the finding that compared to the single time-scale counterpart, the two time-scale algorithm, although with lower complexity, is less robust to noisy feedback, partially due to the sensitivity of the faster time-scale loop to perturbation.

The proofs of our main results make use of a combination of tools in stochastic approximation, Martingale theory and convex analysis. We note that both rate of convergence and two time-scale algorithms for constrained optimization problems are current research topics in stochastic approximation, and indeed our proofs are built on some recent results.

7.1.3. Related work

There has recently been a large number of work on the utility maximization approach for network resource allocation. Most relevant to Section IV of this chapter are those on joint congestion control and MAC (either scheduling-based or random-access-based): e.g., [5, 9, 32, 37, 33, 56, 90, 71, 57, 86, 42, 67, 94, 96].

We now briefly summarize the differences between our work and the most closely related work, on the following two topics that form the core of this study:

Stochastic NUM. There are four levels of stochastic dynamics that need to be brought back into the basic NUM formulations based on deterministic fluid models: session level (sessions come and go with finite workload), packet level (packets arrive in bursts and also go through stochastic marking and dropping), channel level (channel conditions vary), and topology level (network topology varies). Compared to session level stochastic issues such as stability, there has been much less work on packet level stochastic dynamics. Deb, Srikant, and Shakkottai [36] validated fluid approximation for packet level models in the many-user asymptote, and Chang and Liu [29] established the connection between

HTTP layer utility and TCP layer utility. This study investigates packet level stochastic dynamics from the viewpoint of noisy feedback where the "noise" is in part induced by packet level marking and dropping.

We note that, when dealing with the session level randomness where the number of flows changes, stochastic stability means that the number of users and the queue lengths on all the links remain finite [22, 57]. In contrast, in the presence of stochastic perturbations of network parameters, stochastic stability means that the proposed algorithms converge to the optimal solutions in some sense (e.g., almost sure), and this is the subject of our study here.

It is also worth noting that in Kelly's seminal work [52, 51], stochastic stability is examined using linear stochastic perturbation around the equilibrium point, which implicitly assumes the convergence of the distributed algorithms in the first place. Our findings on the rate of convergence provide another step towards understanding how quickly the large time-scale network parameters (e.g. the number of traffic flows) are allowed to fluctuate while the algorithm can still track the optimal point.

NUM with errors. This is an under-explored area with only limited treatment in very few recent publications. The case of deterministic errors in feedback has recently been studied by Mehyar, Spanos, and Low [67], using the methodologies on the gradient method with errors (e.g., [13]). The case of stochastic errors in distributed algorithms for single path congestion control has been investigated in [52], where convergence is assumed and the focus is on the rate of convergence. The multipath case has recently been investigated by Lin and Shroff [58], with focus on the models with unbiased errors on link load measurements for wired network congestion control. In [100], we have taken some preliminary steps to study the impact of stochastic feedback signals on a specific rate control algorithm, assuming that the gradient estimator is unbiased. The impact of noisy feedback on the primal-dual algorithm can be found in the conference version of this work [98].

Alternative decomposition. The conceptually important idea that each generalized NUM may be decomposed in many different ways, each leading to an alternative network architecture, has been realized very recently. For example, alternative decomposition for distributed interactions within one layer has been studied by Palomar and Chiang [74]. Alternative decompositions have been developed over the last year by various researchers for three important classes of NUM: network coding based multicast, joint congestion control and routing in wired networks, and joint congestion control, contention control, scheduling, and routing in wireless multihop networks. The last group of publications is most relevant to this paper, and includes [94, 57, 32, 56, 86, 42, 5].

In addition to providing a totally different decomposition for this problem, we also compare the robustness of alternative decompositions under noisy feedback, which is an important metric to compare alternative decompositions that has not been quantified before.

7.1.4. Organization

The rest of the chapter is organized as follows. Section 7.2 presents the general problem formulation. In Section 7.3, we focus on the distributed NUM algorithm based on Lagrange dual decomposition, and develop a general theory on stochastic stability for distributed P-D algorithms in the presence of noisy feedback, for both the unbiased and biased cases. We also study the rate of convergence for the unbiased case. In Section 7.4,

we apply the general theory summarized above to investigate stability of cross-layer rate

control for joint flow control and random access when strong duality holds. Our numer-

ical examples corroborate the theoretic results well. Section 7.5 contains the study on

stochastic two time-scale algorithms based on primal decomposition. Finally, Section 7.6

concludes the chapter.

7.2. Problem Formulation

7.2.1. Rate Control via Network Utility Utilization (NUM)

Consider a communication network with L links, each with a (possibly time-varying)

capacity of c_l, and S source nodes, each transmitting at a source rate of x_s. We assume

that each source s emits one flow, using a fixed set of links $\mathcal{L}(s)$ in its path, and has a

utility function $U_s(x_s)$. Let $\mathcal{S}$ denote the set of flows in the network, and $\mathcal{S}(l)$ the set of

flows using link l. As is standard, we assume that the network is modelled as a directed

graph $G = (N, E)$, where N is the set of the nodes and E is the set of the directed edges.

The general NUM problem is targeted at maximizing the total utility $\sum_s U_s(x_s)$, over

the source rates $\mathbf{x}$, subject to flow rate constraints for all links l:

$$\Xi_1 : \quad \max_{\{0 \leq x_s \leq M_s\}} \quad \sum_s U_s(x_s)$$
$$\text{subject to} \quad f_l(\mathbf{x}) \leq c_l, \quad \forall l \tag{7.1}$$

where the utilities U_s are twice-differentiable, increasing, and strictly concave functions,

$\{f_l\}$ are twice-differentiable convex functions, M_s is an upper bound on the flow rate,

and c_l is the ergodic capacity of link l, which depends on the specific MAC protocols and

channel conditions.

The above basic NUM has recently been extended to much richer constraint sets involving variables other than just the source rates. In particular, in some network models (e.g, MAC with random access), link capacities c_l can be adapted and the constraints may not be convex.

Built on the above NUM framework, a large number of decentralized algorithms have been devised, with the following general form on primal variables $\mathbf{x}$ and dual variables $\boldsymbol{\lambda}$ (discrete-time version with time index t):

$$x_s(t+1) = F_s(x_s(t), q_s(t)); \quad \forall s \in \mathcal{S} \tag{7.2}$$

$$\lambda_l(t+1) = G_l(y_l(t), \lambda_l(t)), \quad \forall l \in \mathcal{L} \tag{7.3}$$

where F_s and G_l are some non-negative functions, q_s is the overall end-to-end price along the path of flow s, and y_l is the aggregated rate at link l. As surveyed in, e.g., [61, 34], from a reverse engineering perspective, in the case of $\mathbf{x}$ being just the source rates, these algorithms can be viewed as distributed solutions to some optimization problems, and the general algorithm in (7.2) and (7.3) is often implemented through gradient or sub-gradient algorithms.

Roughly speaking, the existing distributed algorithms can be classified into the following three categories (cf. [61]): 1) **Primal-dual algorithm:** In the primal-dual algorithm, the source rates x_s and the shadow prices λ_l are updated on the same time scale. 2) **Primal algorithm:** The primal algorithm often refers to a first-order gradient-descent algorithm for source rate updating and a *static* link price algorithm. 3) **Dual Algorithm:** The dual algorithm is based on dual decomposition, and consists of a gradient-type algorithm for shadow price updating and a *static* source rate algorithm. In general, the primal-dual algorithm has a faster convergence rate.

7.2.2. Distributed NUM Under Noisy Feedback Based on Lagrangian Dual Decomposition

It is clear that to implement the distributed algorithms (7.2) and (7.3), some communications among network elements (e.g., nodes and links), in the form of information feedback (message passing), are critical for computing the gradients or sub-gradients. Unfortunately, in practical systems the feedback is often obtained using error-prone measurement mechanisms and contains random errors. In light of this, we seek to obtain a clear understanding of the conditions under which the general algorithm (7.2) and (7.3), in the presence of noisy feedback, converges to the equilibrium point obtained by deterministic algorithm. In the following, we focus on the distributed algorithms based on Lagrangian dual decomposition. Similar studies can be carried out for algorithms devised using other decomposition methods, and we will elaborate further on this in Section 5.

We now turn to the algorithms based on the Lagrangian dual decomposition. To this end, we first form the Lagrangian of (7.1):

$$L\left(\mathbf{x}, \boldsymbol{\lambda}\right) = \sum_s U_s\left(x_s\right) + \sum_l \lambda_l\left(c_l - f_l(\mathbf{x})\right) \tag{7.4}$$

where $\lambda_l \geq 0$ is the Lagrange multiplier (link price) associated with the linear flow constraint on link l. Then, the Lagrange dual function is

$$Q(\boldsymbol{\lambda}) = \max_{\mathbf{x} \geq 0} \ L(\mathbf{x}, \boldsymbol{\lambda}) \tag{7.5}$$

and the dual problem is given by

$$\mathbf{D} : \min_{\boldsymbol{\lambda} \geq 0} Q(\boldsymbol{\lambda}). \tag{7.6}$$

When strong duality holds, the primal problem can be equivalently solved by solving the dual problem. Let Φ be the set of $\boldsymbol{\lambda}$ that minimizes $Q(\boldsymbol{\lambda})$. We assume that strong duality holds and Φ is non-empty and compact (which holds under mild conditions [99]). In what follows, we shall focus on the impact of noise feedback on the primal-dual algorithms based on the Lagrangian dual method, which are summarized as follows.

- **Deterministic primal-dual algorithm:** In the primal-dual algorithm, the updates of source rates x_s and the link prices λ_l are executed at the same time scale:

$$x_s(n+1) = [x_s(n) + \epsilon_n L_{x_s}(\mathbf{x}(n), \boldsymbol{\lambda}(n))]_{\mathcal{D}} , \quad \forall s \tag{7.7}$$

$$\lambda_l(n+1) = [\lambda_l(n) - \epsilon_n L_{\lambda_l}(\mathbf{x}(n), \boldsymbol{\lambda}(n))]_0^\infty , \quad \forall l \tag{7.8}$$

where $L_{x_s}(\cdot)$ is the gradient of L with respect to x_s and $L_{\lambda_l}(\cdot)$ is the gradient of L with respect to λ_l, and $[\cdot]_{\mathcal{D}}$ stands for the projection onto the feasible set $\mathcal{D}$.

- **Stochastic primal-dual algorithm:** In the presence of noisy feedback, the gradients $L_{x_s}(\cdot)$ and $L_{\lambda_l}(\cdot)$ are stochastic. Let $\hat{L}_{x_s}(\cdot)$ and $\hat{L}_{\lambda_l}(\cdot)$ be the corresponding estimators, and the stochastic version of the primal-dual algorithm can be written as follows:

$$x_s(n+1) = \left[x_s(n) + \epsilon_n \hat{L}_{x_s}(\mathbf{x}(n), \boldsymbol{\lambda}(n))\right]_{\mathcal{D}} , \quad \forall s \tag{7.9}$$

$$\lambda_l(n+1) = \left[\lambda_l(n) - \epsilon_n \hat{L}_{\lambda_l}(\mathbf{x}(n), \boldsymbol{\lambda}(n))\right]_0^\infty , \quad \forall l \tag{7.10}$$

For the sake of convenience, we have used the same step size ϵ_n for both source rate and link price updating. We emphasize that this assumption does not incur any loss of generality. If the step sizes are different, the shadow prices and the utility functions can be re-scaled with no effect on the stability.

7.3. Stochastic Stability of Primal-Dual Algorithm Under Noisy Feedback

As noted above, a main objective of this study is to examine the impact of noisy feedback on the convergence behavior of the distributed algorithms, which boils down to the stochastic stability of network dynamics. There are many notions of stochastic stability. In the presence of stochastic perturbations of network parameters, stochastic stability is used to indicate that the algorithms converge to the optimal solutions in some sense; and this is the subject of our study here. In what follows, we focus on the stability behavior of stochastic P-D algorithm, which is a single time-scale algorithm.

7.3.1. Structure of Stochastic Gradients

The structure of the stochastic gradients plays a critical role in the stability behavior of the distributed NUM algorithm. For convenience, let $\{\mathcal{F}_n\}$ be a sequence of $\sigma-$algebras generated by $\{(\mathbf{x}(i), \boldsymbol{\lambda}(i)), \forall\, i \leq n\}$, $E_n[\cdot]$ denote the conditional expectation $E[\cdot|\mathcal{F}_n]$. We have the following observations on $\hat{L}_{x_s}(\cdot, \cdot)$ and $\hat{L}_{\lambda_l}(\cdot, \cdot)$.

1) **Stochastic gradient** $\hat{L}_{x_s}(\cdot, \cdot)$: Observe that

$$\hat{L}_{x_s}\left(\mathbf{x}(n), \boldsymbol{\lambda}(n)\right) = L_{x_s}\left(\mathbf{x}(n), \boldsymbol{\lambda}(n)\right) + \alpha_s(n) + \zeta_s(n),$$

where $\alpha_s(n)$ is the biased estimation error of $L_{x_s}\left(\mathbf{x}(n), \boldsymbol{\lambda}(n)\right)$, given by

$$\alpha_s(n) \triangleq E_n\left[\hat{L}_{x_s}\left(\mathbf{x}(n), \boldsymbol{\lambda}(n)\right)\right] - L_{x_s}\left(\mathbf{x}(n), \boldsymbol{\lambda}(n)\right), \tag{7.11}$$

and $\zeta_s(n)$ is a martingale difference noise, given by

$$\zeta_s(n) \triangleq \hat{L}_{x_s}\left(\mathbf{x}(n), \boldsymbol{\lambda}(n)\right) - E_n\left[\hat{L}_{x_s}\left(\mathbf{x}(n), \boldsymbol{\lambda}(n)\right)\right]. \tag{7.12}$$

2) **Stochastic gradient** $\hat{L}_{\lambda_l}(\cdot, \cdot)$: Observe that

$$\hat{L}_{\lambda_l}\left(\mathbf{x}(n), \boldsymbol{\lambda}(n)\right) = L_{\lambda_l}\left(\mathbf{x}(n), \boldsymbol{\lambda}(n)\right)) + \beta_l(n) + \xi_l(n),$$

where $\beta_l(n)$ is the biased estimation error of $L_{\lambda_l}\left(\mathbf{x}(n), \boldsymbol{\lambda}(n)\right)$, given by

$$\beta_l(n) \triangleq E_n\left[\hat{L}_{\lambda_l}\left(\mathbf{x}(n), \boldsymbol{\lambda}(n)\right)\right] - L_{\lambda_l}\left(\mathbf{x}(n), \boldsymbol{\lambda}(n)\right), \tag{7.13}$$

and $\xi_l(n)$ is a martingale difference noise, given by

$$\xi_l(n) \triangleq \hat{L}_{\lambda_l}\left(\mathbf{x}(n), \boldsymbol{\lambda}(n)\right) - E_n\left[\hat{L}_{\lambda_l}\left(\mathbf{x}(n), \boldsymbol{\lambda}(n)\right)\right]. \tag{7.14}$$

In practical systems, a popular mechanism for conveying the price information to the source is by packet marking. In particular, there are two standard marking schemes, random exponential marking (REM) [8] and self-normalized additive marking (SAM) [2]; REM is a biased estimator of the overall price along a route, whereas SAM is an unbiased one. Accordingly, $\hat{L}_{x_s}(\cdot, \cdot)$ can be either biased or unbiased. On the other hand, $\hat{L}_{\lambda_l}(\cdot, \cdot)$ can be either unbiased or biased when the aggregated source rate is estimated (e.g., in wireless networks with random access). In Section IV, we provide more examples to illustrate the structure of the stochastic gradients.

7.3.2. The Unbiased Case

In this section, we show that the proposed stochastic primal-dual algorithm converges with probability one to the optimal points, provided that the gradient estimator is asymptotically unbiased. To this end, we impose the following technical conditions:

A1. *We assume that the noise terms in the gradient estimators are independent across iterations.*

A2. *Condition on the step size:* $\epsilon_n > 0$, $\epsilon_n \to 0$, $\sum_n \epsilon_n \to \infty$ and $\sum_n \epsilon_n^2 < \infty$.

A3. *Condition on the biased error:* $\sum_n \epsilon_n |\alpha_s(n)| < \infty$ w.p.1, $\forall$ s *and* $\sum_n \epsilon_n |\beta_l(n)| < \infty$ w.p.1, $\forall$ l.

A4. *Condition on the martingale difference noise:* $\sup_n E_n[\zeta_s(n)^2] < \infty$ w.p.1, $\forall$ s, *and* $\sup_n E_n[\xi_l(n)^2] < \infty$ w.p.1, $\forall$ l.

158

We now present out first main result on the convergence of the stochastic primal-dual algorithm in the presence of unbiased noisy feedback information. The complete proof can be found in Appendix B.1.

Theorem 7.3.1. *Under Conditions* **A1 − A4**, *the iterates* $\{(\mathbf{x}(n), \boldsymbol{\lambda}(n)), n = 1, 2, \ldots\}$, *generated by stochastic algorithms (7.9) and (7.10), converge with probability one to the optimal solution of the primal Problem* $\boldsymbol{\Xi_1}$.

Sketch of the proof: The proof consists of two steps. First, define a stochastic Lyapunov function as the square of the distance between the iterates $(\mathbf{x}, \boldsymbol{\lambda})$ and the set of the saddle points. Via a combination of tools in the Martingale theory and convex analysis, we establish that the iterates generated by (7.9) and (7.10) return to a neighborhood (denoted as A_μ) of the optimal points infinitely often. Then, we show that the recurrent iterates eventually reside in an arbitrary small neighborhood of the optimal points, and the proof involves some tedious local analysis.

Remarks: 1) Clearly, the first key step in the stability analysis is to construct the Lyapunov function. Different from the techniques in [52, 55, Chap. 5], the Lagrangian function in the primal-dual algorithm is neither convex nor concave in $(\mathbf{x}, \boldsymbol{\lambda})$ because the primal problem is a maximization over $\mathbf{x}$ but the dual problem is a minimization over $\boldsymbol{\lambda}$. As a result, the P-D algorithm is much more involved than the standard gradient algorithm [55]. Observing that the ultimate goal is to show that the iterates approach the optimal solutions eventually, we set the distance between the iterates and the set of the optimal saddle points as the Lyapunov function. We elaborate further on this in Appendix A.

2) Condition **A2** is a standard technical condition in stochastic approximation for

proving convergence with probability one (w.p.1), and Condition **A3** essentially requires that the biased term is asymptomatically diminishing. However, if the step size ϵ_n does *not* go to zero (which occurs often in on-line applications) or the biased terms $\{\alpha_s(n), \beta_l(n)\}$ do not diminish (which may be the case in some practical systems), we cannot expect to get probability one convergence to the optimal point. To be quantified in the next subsection, the hope is that the iterates would converge to some neighborhood of the optimal points. It should be cautioned that even the expectation of the limiting process would not be the optimal point if the biased terms do not go to zero.

7.3.3. The Biased Case

As noted above, when the gradient estimators are biased, we cannot hope to obtain almost sure convergence of the stochastic primal-dual algorithm. Nevertheless, if the biased terms are asymptotically bounded by a scaled version ($0 \leq \eta < 1$) of the true gradients, we show that the iterates still move towards the optimal point. For technical reasons, we impose the following condition on the biased terms:

A5. *Condition on the biased error: There exist non-negative constants $\{\alpha_s^u\}$ and $\{\beta_l^u\}$ such that $\limsup_n |\alpha_s(n)| \leq \alpha_s^u$, $\forall s$, and $\limsup_n |\beta_l(n)| \leq \beta_l^u$, $\forall l$, with probability one.*

Define the "contraction region" A_η as follows:

$$A_\eta \triangleq \{(\mathbf{x}, \boldsymbol{\lambda}) : \alpha_s^u \geq \eta |L_{x_s}(\mathbf{x}, \boldsymbol{\lambda})|, \text{ for some } s, \text{ or } \beta_l^u \geq \eta |L_{\lambda_l}(\mathbf{x}, \boldsymbol{\lambda})|, \text{ for some } l, 0 \leq \eta < 1\}.$$

$$(7.15)$$

It is not difficult to see that A_η defines a closed and bounded neighborhood around the equilibrium point, simply because at the equilibrium point $L_{x_s}(\mathbf{x}^*, \boldsymbol{\lambda}^*) = 0$ for all $s \in \mathcal{S}$ and $L_{\lambda_l}(\mathbf{x}^*, \boldsymbol{\lambda}^*) = 0$ for all $l \in \mathcal{L}$, and both $L_{x_s}(\mathbf{x}, \boldsymbol{\lambda})$ and $L_{\lambda_l}(\mathbf{x}, \boldsymbol{\lambda})$ are continuous. The size of A_η depends on the values of η and $\{\alpha_s^u, \beta_l^u\}$. The larger the value of η is, the smaller the size of A_η would be. In contrast, the larger the values of $\{\alpha_s^u, \beta_l^u\}$ are, the larger the size of A_η would be.

Next, we present our second main result on the stochastic stability of the primal-dual algorithm in the presence of biased noisy feedback information.

Theorem 7.3.2. *Under Conditions* **A1**, **A2**, **A4**, *and* **A5**, *the iterates* $\{(\mathbf{x}(n), \boldsymbol{\lambda}(n), n = 1, 2, \ldots\}$, *generated by stochastic algorithms (7.9) and (7.10), return to the set* A_η *infinitely often with probability one.*

The proof is relegated to Appendix B. Note that condition **A5** is weaker than **A3**, and in this sense Theorem 7.3.2 is a generalization of Theorem 7.3.1.

To get a more concrete sense of the regularity conditions, we first observe that outside the set A_η, i.e., in A_η^c,

$$\alpha_s^u \leq \eta |L_{x_s}(\mathbf{x}, \boldsymbol{\lambda})|, \forall \, s,$$

$$\text{and } \beta_l^u \leq \eta |L_{\lambda_l}(\mathbf{x}, \boldsymbol{\lambda})|, \forall \, l.$$

These, combined with Condition A5, indicate that the biased terms are asymptotically uniformly bounded by a scaled version of the true gradient outside A_η. As a result, the errors are relatively small and would not negate the true gradient. Therefore, the inexact gradient can still drive the iterate to move towards the optimal points until it enters A_η.

7.3.4. Rate of Convergence for The Unbiased Case

We now examine the rate of convergence when the stochastic gradients are unbiased. Roughly speaking, the rate of convergence is concerned with the asymptotic behavior of normalized distance of the iterates from the optimal points. Recall that the primal-dual algorithm takes the following general form:

$$
\begin{bmatrix} x_s(n+1) \\ \lambda_l(n+1) \end{bmatrix} = \begin{bmatrix} x_s(n) \\ \lambda_l(n) \end{bmatrix} + \epsilon_n \begin{bmatrix} L_{x_s}(\mathbf{x}(n), \boldsymbol{\lambda}(n)) \\ -L_{\lambda_l}(\mathbf{x}(n), \boldsymbol{\lambda}(n)) \end{bmatrix} + \epsilon_n \begin{bmatrix} \alpha_s(n) + \zeta_s(n) \\ \beta_l(n) + \xi_l(n) \end{bmatrix} + \epsilon_n \begin{bmatrix} Z_n^{x_s} \\ Z_n^{\lambda_l} \end{bmatrix},
$$

$$(7.16)$$

where $\epsilon_n Z_n^{x_s}$ and $\epsilon_n Z_n^{\lambda_l}$ are the reflection terms which "force" x_s and λ_l to reside inside the constraint sets. As is standard in the study on rate of convergence, we focus on local analysis, assuming the iterates generated by the stochastic primal-dual algorithm have entered a small neighborhood of $(\mathbf{x}^*, \boldsymbol{\lambda}^*)$.

To characterize the asymptotic properties, we define $U_{\mathbf{x}}(n) \triangleq (\mathbf{x}(n) - \mathbf{x}^*)/\sqrt{\epsilon_n}$ and $U_{\boldsymbol{\lambda}}(n) \triangleq (\boldsymbol{\lambda}(n) - \boldsymbol{\lambda}^*)/\sqrt{\epsilon_n}$, and we construct $U^n(t)$ to be the piecewise constant interpolation of $U(n) = \{U_{\mathbf{x}}(n), U_{\boldsymbol{\lambda}}(n)\}$, i.e., $U^n(t) = U(n+i)$, for $t \in [t_{n+i} - t_n, t_{n+i+1} - t_n)$, where $t_n \triangleq \sum_{i=0}^{n-1} \epsilon_n$.

A6. Let $\boldsymbol{\theta}(n) \triangleq (\mathbf{x}(n), \boldsymbol{\lambda}(n))$ and $\boldsymbol{\phi}_n \triangleq (\boldsymbol{\zeta}(n), \boldsymbol{\xi}(n))$. Suppose for any given small $\rho > 0$, there exists a positive definite symmetric matrix $\Sigma = \boldsymbol{\sigma}\boldsymbol{\sigma}'$ such that, as $n \to \infty$,

$$
E_n[\boldsymbol{\phi}_n\boldsymbol{\phi}_n^T - \boldsymbol{\Sigma}]I_{\{|\boldsymbol{\theta}(n)-\boldsymbol{\theta}^*|\leq\rho\}} \to 0.
$$

A7. Let $\epsilon_n = 1/n$; and assume that $\mathbf{H} + \mathbf{I}/2$ is a Hurwitz matrix, where $\mathbf{H}$ is the Hessian matrix of the Lagrangian function at $(\mathbf{x}^*, \boldsymbol{\lambda}^*)$.[1]

We have the following proposition on the rate of convergence for the unbiased case.

Proposition 7.3.1. a) Under Conditions **A1, A3, A4, A6** and **A7**, $U^n(\cdot)$ converges in

---

[1]It can be shown that the real parts of the eigenvalues of $\mathbf{H}$ are all non-positive (cf. P. 449 in [13]).

distribution to the solution (denoted as U) to the Skorohod problem:

$$\begin{pmatrix} dU_{\mathbf{x}} \\ dU_{\boldsymbol{\lambda}} \end{pmatrix} = \left(\mathbf{H} + \frac{\mathbf{I}}{2} \right) \begin{pmatrix} U_{\mathbf{x}} \\ U_{\boldsymbol{\lambda}} \end{pmatrix} dt + \boldsymbol{\sigma} dw(t) + \begin{pmatrix} dZ_{\mathbf{x}} \\ dZ_{\boldsymbol{\lambda}} \end{pmatrix}, \qquad (7.17)$$

where $w(t)$ is a standard Wiener process and $Z(\cdot)$ is the reflection term.

b) If $(\mathbf{x}^*, \boldsymbol{\lambda}^*)$ is an interior point of the constraint set, then the limiting process U is a stationary Gaussian diffusion process, and $U(n)$ converges in distribution to a normally distributed random variable with mean zero and covariance $\boldsymbol{\Sigma}$.

c) If $(\mathbf{x}^*, \boldsymbol{\lambda}^*)$ is on the boundary of the constraint set, then the limiting process U is a stationary reflected linear diffusion process.

Sketch of the proof: Proposition 7.3.1 can be proved by combining the proof of Theorem 5.1 in [26] and that of Theorem 2.1 in Chap. 6 in [55]. Roughly, we can expand, via a truncated Taylor series, the interpolated process $U^n(t)$ around the chosen saddle point $(\mathbf{x}^*, \boldsymbol{\lambda}^*)$. Then, a key step needed is to show the tightness of $U^n(t)$. To this end, we can follow Part 3 in the proof of Theorem 2.1 in Chap. 6 in [55] to establish that the biased term in the interpolated process diminishes asymptotically and does not affect the tightness of $U^n(t)$. More specifically, after showing the tightness of $U^n(t)$, one can work with a truncated version of $U^n(t)$ to establish the weak convergence of the truncated processes, and then use the properties of these weak sense limits to show that the truncation does not impact the conclusion. The rest follows from the proof of Theorems 4.3, 4.4 and 5.1 in [26].

Remarks: Proposition 7.3.1 reveals that for the constrained case, the limiting process of the interpolated process for the normalized iterates depends on the specific structure of the Skorohod problem [55]. In general, the limit process is a stationary reflected linear diffusion process, not necessarily a standard Gaussian diffusion process. The limit process would be Gaussian only if there is no reflection term.

Based on Proposition 7.3.1, the rate of convergence depends heavily on the smallest eigenvalue of $\left(\mathbf{H} + \frac{I}{2}\right)$. The more negative the smallest eigenvalue is, the faster the rate of convergence is. Intuitively speaking, the reflection terms help increase the speed of convergence, which unfortunately cannot be characterized exactly. As noted in [26], one cannot readily compute the stationary covariance matrix for reflected diffusion processes, and we have to resort to simulations to get some understanding of the effect of the constraints on the asymptotic variances. Furthermore, the covariance matrix of the limit process gives a measure of the spread at the equilibrium point, which is typically smaller than that of the unconstrained case [26].

Our findings on the rate of convergence provide another step towards understanding how rapidly the slower timescale network parameters (e.g. the number of traffic flows) are allowed to change before the algorithm converges to the optimal point.

7.4. Application to Joint Flow Control and Random Access

In this section, we apply the general theory developed above to investigate the stability of cross-layer rate control for joint flow control and random access, in the presence of noisy feedback. Specifically, we consider rate control in multi-hop wireless networks with random access, and take into account the rate constraints imposed by the MAC/PHY layer. Since random access in wireless networks is node-based, we abuse the notation and denote a link as a pair of nodes (i, j), where i is the transmitter of the link and j is the receiver. For convenience, let $N_{to}^I((i, j))$ denote the set of nodes whose transmissions interfere with link (i,j)'s, excluding node i. Let $N_{in}(i)$ denote the set of the nodes from which node i receives traffic, $N_{out}(i)$ the set of nodes to which node i is sending packets and $L^I_{from}(i)$ to denote the set of links whose transmissions are interfered by node i's transmission, excluding the outgoing links from node i.

We assume that the persistence transmission mechanism is used at the MAC layer, i.e., node i of link (i,j) contends the channel with a persistence probability $p_{(i,j)}$. Define $P_i = \sum_{j \in N_{out}(i)} p_{(i,j)}$. It is known that successful transmission probability of link (i,j) is given by $p_{(i,j)} \prod_{k \in N^I_{to}((i,j))}(1 - P_k)$ [14]. In light of this, we impose the constraint that the total flow rate over link (i,j) should be no more than $r_{(i,j)} p_{(i,j)} \prod_{k \in N^I_{to}((i,j))}(1 - P_k)$, where $r_{(i,j)}$ is the average link rate. For simplicity, in the following, we assume that $\{r_{(i,j)}\}$ is normalized to one.

The cross-layer rate optimization across the transport layer and the MAC layer can be put together as follows:

$$
\mathbf{P}: \quad \max_{\{x_s, p_{(i,j)}\}} \quad \sum_s U_s(x_s)
$$

$$
\text{subject to} \quad \sum_{s \in \mathcal{S}((i,j))} x_s \leq [p_{(i,j)} \prod_{k \in N^I_{to}((i,j))} (1 - P_k)], \forall\, (i,j)
$$

$$
\sum_{j \in N_{out}(i)} p_{(i,j)} = P_i, \; \forall\, i \tag{7.18}
$$

$$
0 \leq x_s \leq M_s, \; \forall\, s
$$

$$
0 \leq P_i \leq 1, \; \forall\, i,
$$

where M_s is the maximum for flow data rate of s, and the utility function $U_s(\cdot)$ is assumed to be [68]:

$$
U_\kappa(x_s) = \begin{cases} w_s \log x_s, & \text{if } \kappa = 1 \\ w_s(1 - \kappa)^{-1} x_s^{1-\kappa}, & \kappa \geq 0, \kappa \neq 1. \end{cases} \tag{7.19}
$$

The above problem is non-convex since the link constraints involves the product term of $\{p_{i,j}\}$. By using a change of variables $\tilde{x}_s = \log(x_s)$ (cf. [94, 56]), however, it can be transformed into the following convex programming problem, provided that $\kappa \geq 1$ (cf. (7.19)):

$$\mathbf{P_1}: \max_{\{\tilde{x}_s, p_{(i,j)}\}} \quad \sum_{s \in \mathcal{S}} U'_s(\tilde{x}_s)$$

$$\text{subject to} \quad \log(\sum_{s \in \mathcal{S}((i,j))} \exp(\tilde{x}_s)) - \log(p_{(i,j)})$$

$$- \sum_{k \in N'_{to}((i,j))} \log(1 - P_k) \leq 0, \ \forall \ (i,j)$$

$$\sum_{j \in N_{out}(i)} p_{(i,j)} = P_i, \ \forall \ i \qquad (7.20)$$

$$-\infty \leq \tilde{x}_s \leq \tilde{M}_s, \ \forall \ s$$

$$0 \leq P_i \leq 1, \ \forall \ i,$$

where $U'_s(\tilde{x}_s) = U_s(\exp(\tilde{x}_s))$ and $\tilde{M}_s = \log(M_s)$. More specifically, it is easy to show that $U'_s(\tilde{x}_s)$ is strictly concave in $\tilde{x}_s$ for utility functions in (7.19) with $\kappa \geq 1$ [56]. Furthermore, since both terms $\log(\sum_{s \in \mathcal{S}((i,j))} \exp(\tilde{x}_s))$ and $-\log(p_{(i,j)}) - \sum_{k \in N'_{to}((i,j))} \log(1 - P_k)$ are convex functions of $\tilde{x}_s$ and $p_{(i,j)}$, it follows that the above problem is strictly convex with a unique optimal point $\mathbf{x}^*$.

Worth noting is that when $0 \leq \kappa < 1$, $U'_s(\tilde{x}_s)$ becomes a convex function. Then the problem boils down to finding the maxima of a convex function over a convex constraint set (no longer a convex program), indicating that the maxima lies on the boundary of the constraint set.

7.4.1. Cross-Layer Rate Control: The $\kappa \geq 1$ Case

In this section, we assume that $\kappa \geq 1$. The Lagrangian function with the Lagrange multipliers $\{\lambda_{(i,j)}\}$ is given as follows:

$$
\begin{aligned}
L(\tilde{\mathbf{x}}, \mathbf{p}, \boldsymbol{\lambda}) \\
= \left\{ \sum_s U'_s(\tilde{x}_s) - \sum_{(i,j)} \lambda_{(i,j)} \log \left(\sum_{s \in \mathcal{S}((i,j))} \exp(\tilde{x}_s) \right) \right\} \\
+ \sum_{(i,j)} \lambda_{(i,j)} \log \left(p_{(i,j)} \prod_{k \in N^I_{to}((i,j))} (1 - P_k) \right).
\end{aligned}
\tag{7.21}
$$

Then, the Lagrange dual function is

$$
Q(\boldsymbol{\lambda}) = \max_{\substack{\sum_{j \in N_{out}(i)} p_{(i,j)} = P_i \\ 0 \leq \mathbf{P} \leq 1 \\ -\infty \leq \tilde{\mathbf{x}} \leq \tilde{\mathbf{M}}}} L(\tilde{\mathbf{x}}, \mathbf{p}, \boldsymbol{\lambda}),
\tag{7.22}
$$

where $\tilde{\mathbf{M}}$ is a vector of $\tilde{M}_s, \forall \ s \in \mathcal{S}$. Thus, the dual problem is given by

$$
\mathbf{D} : \min_{\boldsymbol{\lambda} \geq 0} Q(\boldsymbol{\lambda}).
\tag{7.23}
$$

It is not difficult to see that the Slater condition is satisfied, indicating that there is no duality gap and the strong duality holds. Furthermore, it can be shown that Φ is non-empty and compact [99]. Summarizing, the primal problem can be equivalently solved by solving the dual problem. To this end, we rewrite $Q(\boldsymbol{\lambda})$ as

$$
\begin{aligned}
Q(\boldsymbol{\lambda}) \\
= \max_{\tilde{\mathbf{m}} \leq \tilde{\mathbf{x}} \leq \tilde{\mathbf{M}}} \underbrace{\left\{ \sum_s U'_s(\tilde{x}_s) - \sum_{(i,j)} \lambda_{(i,j)} \log \left(\sum_{s \in \mathcal{S}((i,j))} \exp(\tilde{x}_s) \right) \right\}}_{\triangleq O_{\mathbf{x}}(\boldsymbol{\lambda})} \\
+ \max_{\substack{\sum_{j \in N_{out}(i)} P_{(i,j)} = P_i \\ 0 \leq P_i \leq 1, \ \forall \ i}} \underbrace{\sum_{(i,j)} \lambda_{(i,j)} \log(p_{(i,j)} \prod_{k \in N^I_{to}((i,j))} (1 - P_k))}_{\triangleq O_{\mathbf{p}}(\boldsymbol{\lambda})}
\end{aligned}
$$

which essentially "decomposes" the original utility optimization into two subproblems, i.e., maximizing $O_{\mathbf{x}}(\boldsymbol{\lambda})$ by flow control and maximizing $O_{\mathbf{p}}(\boldsymbol{\lambda})$ via MAC layer random access, which are loosely coupled by the shadow price $\boldsymbol{\lambda}$. The MAC layer problem can then be solved by [56]

$$p_{(i,j)} = \frac{\lambda_{(i,j)}}{\sum_{k \in N_{out}(i)} \lambda_{(i,k)} + \sum_{l:l \in L^l_{from}(i)} \lambda_l}. \tag{7.24}$$

The flow control subproblem can be readily solved by the following gradient method:

$$\tilde{x}_s(n+1) = [\tilde{x}_s(n) + \epsilon_n L_{\tilde{x}_s}(\tilde{\mathbf{x}}(n), \mathbf{p}(n), \boldsymbol{\lambda}(n))]^{\tilde{M}_s}_{-\infty}, \tag{7.25}$$

where ϵ_n is the step size, $[x]^a_b$ stands for $\max(b, \min(a, x))$, and

$$L_{\tilde{x}_s}(\cdot, \cdot, \cdot) \triangleq \dot{U}'_s(\tilde{x}_s(n)) - \exp(\tilde{x}_s(n)) \sum_{(i,j) \in \mathcal{L}(s)} \frac{\lambda_{(i,j)(n)}}{\sum_{k \in \mathcal{S}((i,j))} \exp(\tilde{x}_k(n))}, \tag{7.26}$$

where $\dot{U}'_s(\tilde{x}_s(n))$ is the first order derivative of $U'_s(\tilde{x}_s(n))$.

Next, by using the sub-gradient method, the shadow price updates are given by

$$\lambda_{(i,j)}(n+1) = \left[\lambda_{(i,j)}(n) - \epsilon_n L_{\lambda_{(i,j)}}(\tilde{\mathbf{x}}(n), \mathbf{p}(n), \boldsymbol{\lambda}(n))\right]^{\infty}_0, \tag{7.27}$$

where

$$L_{\lambda_{(i,j)}}(\cdot, \cdot, \cdot) \triangleq \log(p_{(i,j)}(n)) + \sum_{k \in N^l_{to}((i,j))} \log(1 - P_k(n)) - \log\left(\sum_{s \in \mathcal{S}((i,j))} \exp(\tilde{x}_s(n))\right). \tag{7.28}$$

As noted in Section 3, message passing is required to feed back the price information for distributed implementation. In particular, the parameters $\frac{\lambda_{(i,j)}}{\sum_{s \in \mathcal{S}((i,j))} x_s}$ need to be generated at each link in $\mathcal{L}(s)$, and be fed back to the source node along the routing path. It can be seen that the calculation of the ratio $\frac{\lambda_{(i,j)}}{\sum_{s \in \mathcal{S}((i,j))} x_s}$ at node i requires the local information of the shadow price $\lambda_{(i,j)}$ and the total incoming traffic $\sum_{s \in \mathcal{S}((i,j))} x_s$.

Similarly, in equation (7.27), in order to update the shadow price $\lambda_{(i,j)}$, it suffices to have the local information of the total incoming traffic and the two-hop information of $\{P_k, k \in N_{to}^I(j)\}$. Nevertheless, the feedback is noisy and contains possibly random errors. Summarizing, the stochastic primal-dual algorithm for cross-layer rate control, in the presence of noisy feedback, is given as follows:

Stochastic primal-dual algorithm for cross-layer rate control (in what follows, SA stands for stochastic approximation):

- *SA algorithm for source rate updating*:

$$\tilde{x}_s(n+1) = [\tilde{x}_s(n) + \epsilon_n \hat{L}_{\tilde{x}_s}(\tilde{\mathbf{x}}(n), \mathbf{p}(n), \boldsymbol{\lambda}(n))]^{\tilde{M}_s}, \qquad (7.29)$$

where $\hat{L}_{\tilde{x}_s}(\cdot)$ is an estimator of $L_{\tilde{x}_s}(\cdot)$, i.e., the gradient of L with respect to $\tilde{x}_s$.

- *SA algorithm for shadow price updating*:

$$\lambda_{(i,j)}(n+1) = \left[\lambda_{(i,j)}(n) - \epsilon_n \hat{L}_{\lambda_{(i,j)}}(\tilde{\mathbf{x}}(n), \mathbf{p}(n), \boldsymbol{\lambda}(n))\right]^+, \qquad (7.30)$$

where $\hat{L}_{\lambda_{(i,j)}}(\cdot)$ is an estimator of $L_{\lambda_{(i,j)}}(\cdot)$, i.e., the gradient of L with respect to $\lambda_{(i,j)}$.

- *The persistence probabilities are updated by using (7.24)*.

Based on Theorems 7.3.1 and 7.3.2, we have the following result on the stochastic stability of the above algorithm.

Corollary 7.4.1. *(a) (***The unbiased case***) Under Conditions* **A1** − **A4**, *the iterates* $\{(\tilde{\mathbf{x}}(n), \mathbf{P}(n), \boldsymbol{\lambda}(n)), n = 1, 2, \ldots\}$, *generated by stochastic approximation algorithms (7.29), (7.30) and (7.24), converge with probability one to the equilibrium point.*

(b) (**The biased case**) *Under Conditions* **A1 − A2**, **A4** *and* **A5**, *the iterates* $\{(\tilde{\mathbf{x}}(n), \mathbf{P}(n), \boldsymbol{\lambda}(n), n = 1, 2, \ldots\}$, *generated by stochastic approximation algorithms (7.29), (7.30) and (7.24), return to the set* A_η *infinitely often with probability one.*

The proof of part (a) follows essentially along the same line as that of Theorem 7.3.1, except that $G(\cdot)$ depends on $\tilde{\mathbf{x}}(n)$, $\mathbf{p}(n)$ and $\boldsymbol{\lambda}(n)$. More specifically, to show that $G(\cdot) < 0$ for $(\tilde{\mathbf{x}}(n), \boldsymbol{\lambda}(n)) \in A_\mu^c$, we have that

$$
\begin{aligned}
G(\tilde{\mathbf{x}}(n), \boldsymbol{\lambda}(n)) \;\leq\;& L(\tilde{\mathbf{x}}(n), \mathbf{p}(n), \boldsymbol{\lambda}_{\min}^*) - L(\tilde{\mathbf{x}}^*, \mathbf{p}(n), \boldsymbol{\lambda}(n)) \\
=\;& L(\tilde{\mathbf{x}}(n), \mathbf{p}(n), \boldsymbol{\lambda}_{\min}^*) - L(\tilde{\mathbf{x}}^*, \mathbf{p}^*, \boldsymbol{\lambda}_{\min}^*) && (7.31) \\
&+ L(\tilde{\mathbf{x}}^*, \mathbf{p}^*, \boldsymbol{\lambda}_{\min}^*) - L(\tilde{\mathbf{x}}^*, \mathbf{p}^*, \boldsymbol{\lambda}(n)) && (7.32) \\
&+ L(\tilde{\mathbf{x}}^*, \mathbf{p}^*, \boldsymbol{\lambda}(n)) - L(\tilde{\mathbf{x}}^*, \mathbf{p}(n), \boldsymbol{\lambda}(n)). && (7.33)
\end{aligned}
$$

Observing that

1. (7.31) and (7.32) are negative due to the saddle point property of $(\tilde{\mathbf{x}}^*, \mathbf{p}^*, \boldsymbol{\lambda}^*)$, i.e.,

   $$L(\tilde{\mathbf{x}}(n), \mathbf{p}(n), \boldsymbol{\lambda}^*) \leq L(\tilde{\mathbf{x}}^*, \mathbf{p}^*, \boldsymbol{\lambda}^*) \leq L(\tilde{\mathbf{x}}^*, \mathbf{p}^*, \boldsymbol{\lambda}(n));$$

2. (7.33) is negative because when $(\tilde{\mathbf{x}}^*, \boldsymbol{\lambda}(n))$ is fixed, $\mathbf{p}(n)$ is the unique maximum for $L(\cdot, \cdot, \cdot)$;

we conclude that there exists $\delta_\mu > 0$ such that $G(\{\tilde{\mathbf{x}}(n), \boldsymbol{\lambda}(n)\}) < -\delta_\mu$ when $(\tilde{\mathbf{x}}(n), \boldsymbol{\lambda}(n)) \in A_\mu^c$.

The proof of part (b) is a direct application of Theorem 7.3.2.

Example 1: Structure of $\hat{L}_{\tilde{x}_s}(\tilde{\mathbf{x}}(\mathbf{n}), \mathbf{p}(\mathbf{n}), \lambda(\mathbf{n}))$ under random exponential marking (REM):

To get a more concrete sense of Condition **A3** on the biased terms and Condition **A4** on the martingale noise terms, we use the following example to elaborate further on this when random exponential marking is used.

Suppose that the exponential marking technique is used to feedback the price information to the source nodes. More specifically, every link (i, j) marks a packet independently with probability $1 - \exp\left(\frac{-\lambda_{(i,j)}}{\sum_{s \in \mathcal{S}(i,j)} \exp(\tilde{x}_s)}\right)$. Therefore, the end-to-end non-marking probability for flow s is given as follows:

$$q = \exp\left(\sum_{(i,j) \in \mathcal{L}(s)} \frac{-\lambda_{(i,j)}}{\sum_{s \in \mathcal{S}(i,j)} \exp(\tilde{x}_s)}\right)$$

To obtain the estimate of the overall price, source s sends N_n packets during round n and counts the non-marked packets. For example, if K packets have been counted, then the estimation of the overall price can be $-\log(\hat{q})$ where $\hat{q} = K/N_n$. Therefore,

$$\hat{L}_{\tilde{x}_s}(\tilde{\mathbf{x}}(n), \mathbf{p}(n), \boldsymbol{\lambda}(n)) = \dot{U}'_s(\tilde{x}_s(n)) + \exp(\tilde{x}_s(n)) \log(\hat{q}). \tag{7.34}$$

By the definition of (7.11), we have that

$$
\begin{aligned}
\alpha_n &= E_n\left[\hat{L}_{\tilde{x}_s}(\tilde{\mathbf{x}}(n), \mathbf{p}(n), \boldsymbol{\lambda}(n))\right] - L_{\tilde{x}_s}(\tilde{\mathbf{x}}(n), \mathbf{p}(n), \boldsymbol{\lambda}(n)), \\
&= \exp(\tilde{x}_s(n))\left(E_n[\log(\hat{q})] - \log(q)\right)
\end{aligned}
\tag{7.35}
$$

Using Jensen's inequality, it is clear that $\alpha_s(n) \leq 0$ for any s and n. Next, we find an upper-bound for $|\alpha_s(n)|$. Note that K is a Binomial random variable with distribution $B(N_n, q)$. When N_n is sufficiently large, it follows that $\hat{q} \sim \mathcal{N}(q, q(1-q)/N_n)$ and $\hat{q} \in [q - c/\sqrt{N_n}, q + c/\sqrt{N_n}]$ with high probability, where c is some positive constant (great than 3). That is, the estimation bias of the overall shadow price is upper-bounded by

$$|\alpha_s(n)| \leq \tilde{M}_s |E[\log(\hat{q})] - \log(q)| \leq \frac{c'_s}{\sqrt{N_n}} \tag{7.36}$$

for large N_n, where c'_s is some positive constant.

Clearly, if a fixed number N_n of packets for round n is used, the biased term would exist. In contrast, to meet Condition **A3** such that the biased term should diminish asymptotically, it suffices to have that

$$\sum_n \frac{\epsilon_n}{\sqrt{N_n}} < \infty. \tag{7.37}$$

For example, when $\epsilon_n = 1/n$, $N_n \sim O(\log^4(n))$ would satisfy (7.37), indicating that it suffices to have the measurement window size grows at the rate of $\log^4(n)$.

Next, we examine the variance of $\zeta_s(n)$ to check if Condition **A4** is satisfied. By (7.12) and (7.34), it follows that

$$
\begin{aligned}
E_n[\zeta_s(n)^2] &= E_n\left[\hat{L}_{\tilde{x}_s}\left(\tilde{\mathbf{x}}(n), \mathbf{p}(n), \boldsymbol{\lambda}(n)\right) - E_n\left[\hat{L}_{\tilde{x}_s}\left(\tilde{\mathbf{x}}(n), \mathbf{p}(n), \boldsymbol{\lambda}(n)\right)\right]\right]^2 \\
&= E_n\left[\exp(\tilde{x}_s(n))\left(\log(\hat{q}) - E_n[\log(\hat{q})]\right)\right]^2 \\
&\leq \tilde{M}_s^2 E_n[\log(\hat{q})]^2 \\
&\leq \tilde{M}_s^2 E_n[\log(q+c)]^2.
\end{aligned}
$$

Example 2: Structure of $\hat{L}_{\lambda_{(i,j)}}(\tilde{\mathbf{x}}(n), \mathbf{p}(n), \boldsymbol{\lambda}(n))$ under random access:

To update the shadow prices, each link needs to estimate the incoming flow rate $t = \sum_{s \in \mathcal{S}((i,j))} \exp(\tilde{x}_s(n))$ (or equivalently $\sum_{s \in \mathcal{S}((i,j))} x_s(n)$). Since the packets are transmitted using random access, there can be at most one successful transmission at each time slot. Therefore, the packet arrival rate at link (i,j) at each time slot follow a Bernoulli distribution with a successful probability $\sum_{s \in \mathcal{S}((i,j))} x_s$. The number of arrived packets K during a time window of M_n can be used to estimate the rate, and the estimator $\hat{t} = K/M_n$ follows a normal distribution $\mathcal{N}(t, t(1-t)/M_n)$ as M_n is reasonably large. To

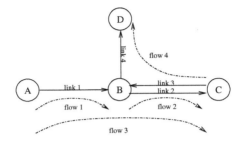

Figure 7.1. A simple network model.

get an upper bound for $\beta_{(i,j)}(n)$, it is easy to see from (7.13) and (7.27) that

$$|\beta_{(i,j)}| \leq |E[\log(\hat{t})] - log(t)| \leq \frac{e_{(i,j)}}{\sqrt{M_n}}, \tag{7.38}$$

for large M_n, where $e_{(i,j)}$ is a positive constant. Thus, if M_n satisfies the same condition

as N_n in (7.37), then Condition **A3** is met. Similar studies can be carried out to $\xi_{(i,j)}(n)$

in (7.14).

7.4.2. Numerical Examples

In this section, we illustrate, via numerical examples, the convergence performance

of the single time-scale algorithm. Specifically, we consider a basic network scenario, as

depicted in Fig. 7.1. There are four nodes (A, B, C and D), four links and four flows,

where flow 1 is from node A to node B, flow 2 from B and to C, flow 3 from A to C via

B, and flow 4 from C to D via B. The utility function in Problem Ξ is taken to be the

logarithm utility function where $\kappa = 1$, $w_s = 1$, and $M_s = 1, \forall\ s$.

Table 7.1. Comparison between the result of the proposed primal-dual algorithm and the optimal solution.

Link probabilities	$p_{(A,B)}$	$p_{(B,C)}$	$p_{(C,B)}$	$p_{(B,D)}$
Primal-dual algorithm	0.6655	0.3840	0.2009	0.0391
Optimal solution	0.6457	0.3750	0.2152	0.0443
Flow rate	x_1	x_2	x_3	x_4
Primal-dual algorithm	0.2065	0.2065	0.0974	0.0385
Optimal solution	0.1962	0.1962	0.0981	0.0443

7.4.2.1. Convergence of The Deterministic Algorithm

We first assume that there is no estimation error for the gradients. The results obtained by the primal-dual algorithm, together with the theoretical optimal solution, are summarized in Table 7.1. It can be seen that the result obtained from the primal-dual algorithm corroborates the optimal solutions. Moreover, Fig. 7.2 illustrates the convergence rate of the primal-dual algorithm under different choices of the step size ϵ_n: Case I with a fixed value $\epsilon_n = 0.01$ and Case II with diminishing step sizes $\{\epsilon_n = 1/n\}$. Clearly, the iterates using step size $\epsilon_n = 0.01$ converge much faster than that using $\epsilon_n = 1/n$. We should caution that this does not necessarily indicate that using constant step sizes is better than diminishing step sizes, particularly in the presence of noisy feedback, and this is the subject of the next section.

7.4.2.2. Stability of The Stochastic Algorithm: The Unbiased Case

Next, we examine the convergence of the primal-dual algorithm under noisy feedback. Note that since the observation time window size has many samples, the random noise can be well approximated by the Gaussian random variable. Therefore, for ease of exposition, we let $\alpha_s(n) = \beta_{(i,j)}(n) = 1/n$, $\zeta_s(n) \sim \mathcal{N}(0,1)$ and $\xi_{(i,j)}(n) \sim \mathcal{N}(0,1)$, for all s and (i,j).

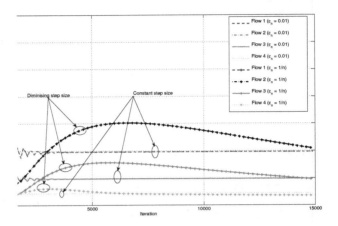

Figure 7.2. Convergence of deterministic algorithm under different step sizes.

Fig. 7.3 depicts the sequence of the iterates produced by the primal-dual algorithm under different ϵ_n. Comparing Fig. 7.3 and Fig. 7.2, we observe that the iterates using constant step sizes are less robust to the random perturbation, in the sense that the fluctuation of the iterates remains significant after many iterations; in contrast, the fluctuation using diminishing step sizes "vanishes" eventually. This corroborates Theorems 7.3.1 and 7.3.2 which reveal that with constant step sizes, the convergence to a neighborhood is the best we can hope; whereas by using diminishing step sizes, convergence with probability one to the optimal points is made possible.

7.4.2.3. Stability of The Stochastic Algorithm: The Biased Case

Recall that when the gradient estimation error is biased, we cannot hope to obtain almost sure convergence to the optimal solutions. Instead, we have shown that provided

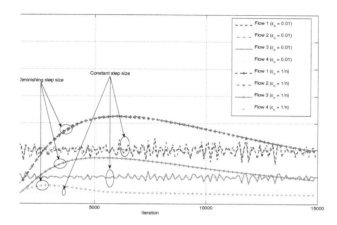

Figure 7.3. Convergence under noisy feedback (the unbiased case).

that the biased error is asymptotically uniformly bounded, the iterates return to a "contraction region" infinitely often. In this example, we assume that $\alpha_s(n) = \beta_{(i,j)}(n)$ and are uniformly bounded by a specified positive value. We also assume that $\zeta_s(n) \sim \mathcal{N}(0,1)$ and $\xi_{(i,j)}(n) \sim \mathcal{N}(0,1)$, for all s and (i,j).

We plot the iterates (using the relative distance to the optimal points) in Fig. 7.4, which is further "zoomed in" in Fig. 7.5. It can be observed from Fig. 7.4 that when the upper-bounds on the $\{\alpha_s, \beta_{(i,j)}\}$ are small, the iterates return to a neighborhood of the optimal solution. However, when the estimation errors are large, the recurrent behavior of the iterates may not occur, and the iterates may diverge. This corroborates the theoretical analysis. We can further observe from Fig. 7.5 that the smaller the upper-bound is, the smaller the "contraction region" A_η becomes, indicating that the iterates converge "closer" to the optimal points.

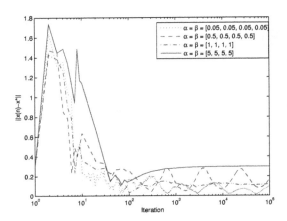

Figure 7.4. Convergence under noisy feedback (the biased case).

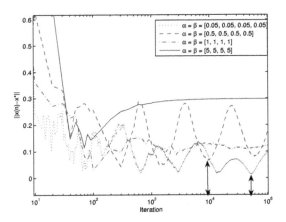

Figure 7.5. "Zoomed-in" convergence behavior of the iterates in Figure 7.4.

7.5. Stochastic Stability of Two Time-Scale Algorithm Under Noisy Feedback

In the previous sections, we have applied the dual decomposition method to Problem (7.1) and devised the primal-dual algorithm, which is a single time-scale algorithm. As noted in Section 7.1, there are many other decomposition methods. In particular, the primal decomposition method is a useful machinery for problem with coupled variables [75]; and when some of the variables are fixed, the rest of the problem may decouple into several subproblems. This naturally yields multiple time-scale algorithms. It is also of great interest to examine the stability of the multiple time-scale algorithms in the presence of noisy feedback, and compare with the single time-scale algorithms, in terms of complexity and robustness.

To get a more concrete sense of the two time-scale algorithms based on primal decomposition, we consider the following NUM problem:

$$\Xi_2: \quad \underset{\{m_s \le x_s \le M_s, \ \mathbf{p}\}}{\text{maximize}} \quad \sum_s U_s(x_s)$$
$$\text{subject to} \quad \sum_{s:l \in \mathcal{L}(s)} x_s \le c_l, \quad \forall l \quad (7.39)$$
$$c_l = h_l(\mathbf{p}), \quad \forall l$$
$$\mathbf{p} \in \mathcal{H},$$

where the link capacities $\{c_l\}$ are functions of specific MAC parameters $\mathbf{p}$ (for instance, $\mathbf{p}$ can be transmission probabilities or powers), i.e., $c_l = h_l(\mathbf{p})$ with $h_l : \mathcal{R}^L \rightarrow \mathcal{R}$, $\{m_s\}$ and $\{M_s\}$ are positive lower and upper bounds on the flow rates, and $\mathcal{H}$ is the constraint set of the MAC parameters $\mathbf{p}$. Since the focus of this paper is on the impact of noisy feedback rather than that of nonconvexity, we assume that the above problem Ξ_2 is a convex optimization (possibly after some transformation).

The deterministic two time-scale algorithm based on primal decomposition for solving

(7.39) is given as follows:

- **Faster (smaller) time scale:**

  - *The source rates are updated by*

  $$x_s(n) = \underset{m_s \leq x_s \leq M_s}{\arg\max} \left(U_s(x_s) - x_s \sum_{l \in \mathcal{L}(s)} \lambda_l(n) \right), \quad \forall s. \qquad (7.40)$$

  - *The shadow prices are updated by*

  $$\lambda_l(n+1) = \left[\lambda_l(n) - a_n \left(c_l(\mathbf{p}(n)) - \sum_{s \in \mathcal{S}(l)} x_s(n) \right) \right]_0^\infty, \quad \forall l, \qquad (7.41)$$

  where a_n is the positive step size.

- **Slower (Larger) time scale:**

  - *The MAC parameters are updated by*

  $$p_l(n+1) = \left[p_l(n) + b_n \sum_{k \in E} \lambda_k^* \frac{\partial h_k(\mathbf{p}(n))}{\partial p_l} \right]_{\mathcal{H}}, \quad \forall l, \qquad (7.42)$$

  where b_n is the positive step size and λ^* is the optimal shadow price generated by the faster time-scale iteration when it converges.

Remarks: 1) Note that in the P-D algorithm based on the dual decomposition, local message passing is needed to update the persistence probabilities (cf. (7.24)) every iteration. However, in the two time-scale algorithm, the updating of the persistence probabilities is performed less frequently than that of the source rates and shadow prices. Therefore, message passing is significantly reduced in the two time-scale algorithm based on primal decomposition. Accordingly, the two time-scale algorithm has a lower communication complexity.

2) There are many ways to implement the above two time-scale algorithm. One method is to run the two loops sequentially, i.e., let the faster time-scale iteration (7.40) and (7.41) converge first, and then execute the slower time-scale loop. Alteratively, one can set $b_n = o(a_n)$ and run the algorithms at both time scales. A plausible way to implement this is to execute one update on the slower time-scale loop for every few iterations on the faster time-scale loop, i.e., the slower time-scale loop is run less frequently. Needless to say, different implementation methods have different complexity and robustness issues. In the following, we focus on the second method for implementing the two time-scale algorithm, and examine its stability performance under noisy feedback.

7.5.1. Stability of The Stochastic Two Time-Scale Algorithm

Next, we examine the impact of the noisy feedback on the stability of the following stochastic two time-scale algorithm based on the primal decomposition method:

- **Faster time scale:**

  - *SA algorithm for source rate updating:*

  $$x_s(n) = \arg\max_{m_s \leq x_s \leq M_s} \left(U_s(x_s) - x_s \left[\sum_{l \in \mathcal{L}(s)} \lambda_l(n) + M_s^x(n) \right] \right). \qquad (7.43)$$

  where $\mathbf{M}^x(n)$ is the estimation error of the end-to-end price (possibly consisting of a bias term).

  - *SA algorithm for shadow price updating:*

  $$\lambda_l(n+1) = \left[\lambda_l(n) - a_n \left\{ \left(c_l(\mathbf{p}(n)) - \sum_{s \in \mathcal{S}(l)} x_s(n) \right) + M_l^\lambda(n) \right\} \right]_0^\infty, \qquad (7.44)$$

  where $\mathbf{M}^\lambda(n)$ is the estimation error of the flow rate (possibly consisting of a bias term).

- **Slower time scale:**

    - *SA algorithm for MAC parameter updating:*

$$p_l(n+1) = \left[p_l(n) + b_n \left(\sum_{k \in E} \lambda_k(n+1) \frac{\partial h_k(\mathbf{p}(n))}{\partial p_l} + N_l(n) \right) \right]_{\mathcal{H}}, \qquad (7.45)$$

where $\mathbf{N}(n)$ is the estimation error of the sub-gradients (possibly consisting of a bias term).

Note that $\mathbf{M}^x(n)$, $\mathbf{M}^\lambda(n)$ and $\mathbf{N}(n)$ are $\mathcal{F}(n)$-measurable, where $\mathcal{F}(n)$ is the σ-algebra generated by the observations made up to time n.

7.5.2. Stability For The Unbiased Case

Define $t_n^a \triangleq \sum_{i=0}^{n-1} a_i$, and $m^a(t) \triangleq m$ such that $t_m^a \le t < t_{m+1}^a$. Define t_n^b and $m^b(t)$ analogously with b_i in lieu of a_i. Define

$$\mathbf{M}^{a,x}(t) \triangleq \sum_{i=0}^{m^a(t)-1} a_i \mathbf{M}^x(i), \qquad (7.46)$$

$$\mathbf{M}^{a,\lambda}(t) \triangleq \sum_{i=0}^{m^a(t)-1} a_i \mathbf{M}^\lambda(i), \qquad (7.47)$$

$$\mathbf{N}^b(t) \triangleq \sum_{i=0}^{m^b(t)-1} b_i \mathbf{N}(i). \qquad (7.48)$$

In order to establish the convergence of the above stochastic two time-scale algorithm in the presence of unbiased error, we impose the following assumptions:

B1. *Condition on the step sizes:*

$$a_n > 0, \ b_n > 0,$$

$$\sum_n a_n = \sum_n b_n = \infty,$$

$$a_n \longrightarrow 0, b_n \longrightarrow 0, b_n/a_n \longrightarrow 0.$$

B2. *Condition on the noises: For some positive T,*

$$\limsup_n \max_{j \geq n} \max_{0 \leq t \leq T} |\mathbf{M}^{a,x}(jT + t) - \mathbf{M}^{a,x}(jT)| = 0 \quad w.p.1,$$

$$\limsup_n \max_{j \geq n} \max_{0 \leq t \leq T} |\mathbf{M}^{a,\lambda}(jT + t) - \mathbf{M}^{a,x}(jT)| = 0 \quad w.p.1,$$

$$\limsup_n \max_{j \geq n} \max_{0 \leq t \leq T} |\mathbf{N}^b(jT + t) - \mathbf{N}^b(jT)| = 0 \quad w.p.1.$$

B3. *Condition on the utility functions: The curvatures of U_s are bounded: $0 < 1/\bar{\nu}_s \leq -U_s''(x_s) \leq 1/\bar{\mu}_s < \infty$ for all $x_s \in [m_s, M_s]$.*

Recall that Problem Ξ_2 in (7.39) is convex. We have the following result.

Theorem 7.5.3. *Under Conditions $\mathbf{B1 - B3}$, the iterates $\{(\mathbf{x}(n), \lambda(n), \mathbf{p}(n)), n = 1, 2, \ldots\}$ generated by stochastic approximation algorithms (7.43), (7.44) and (7.45) converge with probability one to the optimal solutions of Problem Ξ_2.*

Sketch of the proof: A key step in this proof of is to establish the passage from the stochastic two time-scale algorithm to the mean ODEs at two time scales. Then the rest of the proof follows the convergence of the mean ODEs [23, 55]. The details of the proof are presented in Appendix C.

7.5.3. Stability For The Biased Case

The estimation error can also be biased in the two time-scale algorithm, i.e., there is no guarantee that $|\mathbf{M}^x(n)| \to 0$, $|\mathbf{M}^\lambda(n)| \to 0$ and/or $|\mathbf{N}(n)| \to 0$ with probability one. The convergence of the two time-scale algorithm is highly non-trivial in these cases.

We caution that besides the possible bias in gradient estimation, there may exist another cause of the biased error in the two time-scale algorithm. Recall that one popular approach is to let the faster time-scale iteration "converge" first, and then execute the slower time-scale iteration. Since the "convergence" of the faster time-scale iteration is determined by comparing the distance of the iterates with a pre-set threshold, an artificial biased error may be induced for the slower time-scale iteration.

For the case when $\mathbf{M}^x(n)$ and $\mathbf{M}^\lambda(n)$ are asymptotically unbiased but $\mathbf{N}(n)$ is biased, one can construct a contraction region, using the same method as in the stochastic P-D algorithm. However, when $\mathbf{M}^x(n)$ and $\mathbf{M}^\lambda(n)$ are biased, it is very challenging to define/examine the contraction region for the two time-scale algorithm, because the impact of the cumulative effect of the biases on slower time scale iteration can vary significantly.

7.5.4. Numerical Examples

We consider the same network model in Section 7.4.2. The utility functions are the logarithm function with $w_s = 1$, $m_s = 1$ and $M_s = 50$. The link capacity $r_{i,j}$ is scaled to be 50 for all i, j for numerical purposes. Therefore, the corresponding optimal flow rate is $\mathbf{x}^* = [9.810, 9.810, 4.900, 2.215]$ while the optimal link persistence probabilities remain the same as before. The step sizes are $a_n = \frac{0.01}{\lceil n^{0.75} * 500 \rceil}$ and $b_n = \frac{0.001}{\lceil n * 500 \rceil}$.

We first examine the convergence performance of the two time-scale algorithm (7.43), (7.44) and (7.45) when the noisy feedback is unbiased. The estimation noises $\{M_s^x(n)\}$, $\{M_l^\lambda(n)\}$ and $\{N_l(n)\}$ are set to be zero-mean Gaussian random variables, and the standard deviations of $\{M_l^\lambda(n)\}$ and $\{N_l(n)\}$ are fixed at 0.001 while the standard deviation of $\{M_s^x(n)\}$ varies from 0.001 to 0.01. We should point out that the variance of the gradient estimation error in (7.44) is significantly larger than that of $\{M^x(n)\}$, $\{M_l^\lambda(n)\}$ and $\{N_l(n)\}$, due to the nonlinear transformation of $\{M^x(n)\}$ in (7.44).

Fig. 7.6 and Fig. 7.7 depict the simulation results under different noise levels. Comparing Fig. 7.6 and Fig. 7.7, we observe that when the standard deviation of $\{M_s^x(n)\}$ increases, the faster time-scale iteration is more susceptible to the random perturbation than the slower time-scale one. Our intuition is as follows:

- The step size of the slower time-scale iteration is smaller than that of the faster time-scale one, and a smaller step size can help to reduce the noise effect;

- Due to nonlinear transformation, the noise $\{M_s^x(n)\}$ in (7.43) is amplified many folds in (7.44), and thus the induced noise in (7.44) is more significant. More specifically, the "amplified" noise due to $\{M_s^x(n)\}$ can be calculated as

$$\sum_{s\in S(l)} \frac{1}{\sum_{l\in L(s)} \lambda_l + M_s^x} - \sum_{s\in S(l)} \frac{1}{\sum_{l\in L(s)} \lambda_l}$$

$$\approx \sum_{s\in Sl} \frac{1}{\left(\sum_{l\in L(s)} \lambda_l\right)^2} M_s^x$$

$$\approx \sum_{s\in S(l)} x_s^2 M_s^x. \tag{7.49}$$

It follows that, around the equilibrium points, the "amplified" noises in (7.44) due to $M_s^x(n)$ are x_s^2 times larger, which amounts to 100 times for flow 1 and flow 2, 25 times for flow 3 and 5 times for flow 4. This also indicates that the impact of random perturbation on flows with higher rates are higher than that on flows with lower rates, which can be observed in Fig.7.7.

We note that when the standard deviation of $\{M_s^x(n)\}$ are further increased to 0.05, the fluctuation of the faster time-scale iteration is so significant that it may drive the two

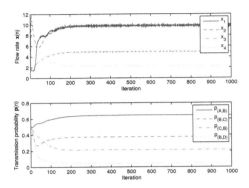

Figure 7.6. Convergence of the two time-scale algorithm (standard deviation of $M_s^x(n)$ is 0.001)

time-scale algorithm to diverge (since $-U_s''(x_s) = 1/x_s^2$ is large when x_s is very small, and Condition **B3** may got violated). In contrast, increasing the standard deviations of $\{M_l^\lambda\}$ or $\{N_l\}$ does not impact the convergence of the two time-scale algorithm much (see Fig. 7.9). This reveals that the noise term $\{M_s^x(n)\}$ plays an important role in the overall stability of the two time-scale algorithm.

Summarizing, our findings reveal that the faster time-scale iteration is more "vulnerable" to noisy perturbation than the slower time-scale one, and its behavior is closely tied to in the stability of the two time-scale algorithm. Consequently, we conclude that the two time-scale algorithm is less robust to the noisy feedback compared with the single time-scale algorithm.

Finally, we examine the performance of the two time-scale algorithm under biased noises. Particularly, we set the noises as follows: $M_s^x(n) = \mathcal{N}(0.1, 10^{-6})$, $\forall\, s, n$, $M_l^\lambda(n) = \mathcal{N}(0.1, 10^{-2})$, $\forall\, l, n$ and $N_l(n) = \mathcal{N}(1, 1)$, $\forall\, l, n$. It can be seen from Fig. 7.10 and

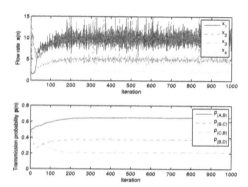

Figure 7.7. Convergence of the two time-scale algorithm (standard deviation of $M_s^x(n)$ is 0.01)

Fig. 7.11 that the iterates converge to some point in the neighborhood of the optimal point. We speculate that there may exist a contraction region for the two time-scale algorithm under appropriate conditions on the biased noise terms, and more work in this avenue is needed to obtain a thorough understanding.

7.6. Conclusion

Stochastic noisy feedback is a practically inevitable phenomenon that has not been systematically investigated in the vast recent literature on distributed NUM. Considering the general distributed algorithms based on different decomposition methods, we make use of a combination of tools in stochastic approximation, Martingale theory and convex analysis to investigate the impact of stochastic noisy feedback:

1. For the unbiased gradient case, we have established that the iterates, generated by the distributed P-D algorithm based on Lagrange dual method, converge with

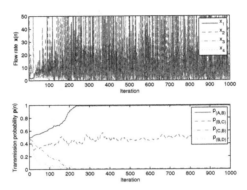

Figure 7.8. Convergence of the two time-scale algorithm with standard deviation of $M_s^x(n)$ set to be 0.05.

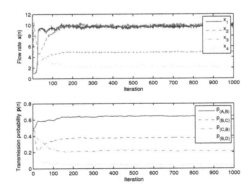

Figure 7.9. Convergence of the two time-scale algorithm with standard deviations of $M_s^x(n)$, $M_l^\lambda(n)$ and $N_l(n)$ set to be 0.001, 0.1 and 1 respectively.

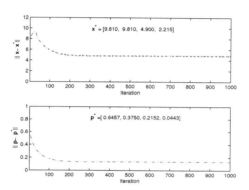

Figure 7.10. Distance of the iterates to the optimal solution under biased feedback.

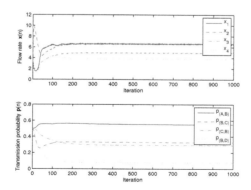

Figure 7.11. Convergence under biased feedback.

probability one to the optimal solution of the centralized algorithm.

2. In contrast, when the gradient estimator is biased, we have showed that the P-D algorithm converges to a contraction region around the optimal point, provided that the biased terms are asymptotically (uniformly) bounded by a scaled version ($0 \leq \eta < 1$) of the true gradients.

3. We have also studied the rate of convergence of the P-D algorithm for the unbiased case, and our results reveal that in general, the limit process of the interpolated process for the normalized iterate sequence is a stationary reflected linear diffusion process, not necessarily a Gaussian diffusion process.

4. Finally, we have investigated stochastic NUM solutions based on alternative decomposition methods. Specifically, we have studied the impact of noisy feedback on the stability of the two time-scale algorithm based on primal decomposition. Using the mean ODE method, we show that the convergence with probability one can be assured under appropriate conditions. Compared with the single time-scale dual algorithm, the two time-scale algorithm has low complexity, but is less robust to noisy feedback, partially due to the sensitivity of the faster time-scale loop to perturbation.

CHAPTER 8

CONCLUSION AND FUTURE DIRECTION

We studied physical-layer aware control and optimization in stochastic wireless networks. In particular, this dissertation is composed of two main thrusts: Thrust I is
targeted at developing a framework for channel-aware distributed scheduling to exploit
rich PHY/MAC diversities; Thrust II is focused on optimization and control for random
access and flow control in wireless ad hoc networks using a stochastic network utility
maximization approach.

8.1. Channel-aware Distributed Scheduling

In Chapters 2-4, we studied distributed opportunistic scheduling to exploit rich
PHY/MAC diversities. A key observation is that in ad-hoc communications, distributed
opportunistic scheduling boils down to a process of joint channel probing and distributed
scheduling. We first investigated distributed opportunistic scheduling from a network-
centric point of view, where links cooperate to maximize the overall network throughput.
We formulated the channel probing and scheduling as a maximal-rate-of-return problem,
and characterized the optimal strategy in the sense of maximizing the overall throughput,
for both homogenous networks and heterogeneous networks. We also investigated DOS
under fairness constraints and time constraints. We showed that the optimal strategy
also has a threshold structure, where the threshold is the solution to a fixed point equation. We then generalized DOS to take into account the noisy channel estimation. For
such cases, we proposed that the transmitted rate backs off on the estimated rate so as
to reduce the channel outage probability. We showed that the optimal scheduling policy is still threshold-based, but the threshold turns out to be a function of the variance

of the estimation error, and is a functional of the backoff rate. Next, we studied channel aware distributed scheduling for exploiting multi-channel/multi-receiver diversities. Specifically, we considered four probing mechanisms in phase II, namely, 1) random selection, 2) exhaustive sequential probing with recall, 3) sequential probing without recall, and 4) sequential probing with recall. We derived the corresponding optimal scheduling policies, all of which exhibit threshold structures.

In Chapter 5, we explored practical MAC design for multi-channel wireless networks. Specifically, we proposed an opportunistic multi-channel medium access control protocol (OMC-MAC) based on the IEEE 802.11 standards. More specifically, the OMC-MAC exploits the channel variations among multiple channels and selects the best channel for data communications, resulting in multi-channel diversity. A key feature of OMC-MAC is that the RTS/CTS/RES handshake is used not only for reserving the communication floor, but also for channel allocation and rate adaptation purposes. Furthermore, we proposed to use a sequential Monte Carlo method, namely, the Sampling Importance Resampling particle filter to estimate the number of competing stations, and adjust the contention window size accordingly.

8.2. Utility-based Random Access and Flow Control

In Chapter 6, we proposed a two-phase utility maximization approach to study wireless MAC design towards QoS provisioning. Based on the observation that the topology variation and channel variation occur on different time scales, the utility optimization is "decoupled" into two phases, i.e., the global phase and the local phase. More specifically, in the global phase, the adaptive persistence mechanism is used to achieve the long-term fairness; and in the local phase, the transmission duration is adapted based on the channel condition and the QoS constraints, aiming to achieve the short-term fairness. Under this two-phase utility maximization framework, we presented the detailed algorithm for the weighted proportional fair MAC. In particular, we utilized a stochastic approximation method to design the adaptive persistence mechanism for the global phase, and establish the stability and analyze the fairness therein.

In Chapter 7, we investigated stochastic distributed NUM algorithm for flow rate control and random access. We observed that stochastic noisy feedback is a practically inevitable phenomenon that has not been systematically investigated in the vast recent literature on distributed NUM. Using a combination of tools in stochastic approximation, Martingale theory and convex analysis, we have identified the conditions for the stability of the distributed NUM algorithms under the impact of stochastic noisy feedback. Specifically, for the unbiased gradient case, we have established that the iterates, generated by the distributed P-D algorithm based on Lagrange dual method, converge with probability one to the optimal solution of the centralized algorithm. In contrast, when the gradient estimator is biased, we have showed that the P-D algorithm converges to a contraction region around the optimal point, provided that the biased terms are asymptotically (uniformly) bounded by a scaled version ($0 \leq \eta < 1$) of the true gradients. We have also studied the rate of convergence of the P-D algorithm for the unbiased case, and our results reveal that in general, the limit process of the interpolated process for the normalized iterate sequence is a stationary reflected linear diffusion process, not necessarily a Gaussian diffusion process. Finally, we have investigated stochastic NUM solutions based on alternative decomposition methods. Specifically, we have studied the impact of noisy feedback on the stability of the two time-scale algorithm based on primal decomposition. Using the mean ODE method, we show that the convergence with probability one can be assured under appropriate conditions. Compared with the single time-scale dual algorithm, the two time-scale algorithm has low complexity, but is less robust to noisy feedback, partially due to the sensitivity of the faster time-scale loop to perturbation.

8.3. Future Work

8.3.1. Channel-aware Distributed Scheduling

We have studied channel-aware distributed scheduling from a network-centric perspective where users cooperate to maximize the overall network throughput. There are a number of directions deserving further investigations.

1. It is of great interest to consider distributed opportunistic scheduling from a user-perspective where users compete for the channel resources to maximize their individual throughput in a selfish manner.

2. We have considered distributed opportunistic scheduling to address fairness constraints, where the objective is to maximize the expected value of the utility functions. It is of equal importance to consider maximizing the sum of the utilities of the average throughput for each individual users. We expect that such studies can be carried out from a sample path argument [86, 55].

3. We have considered distributed opportunistic scheduling under time constraints, where the constraint is on the average probing time. This is suitable for time-sensitive applications. For real time traffic, a hard deadline on the probing time

would be more appropriate. Again, a sample-path based solution is useful to tackle this type of problem. Ultimately, an outage constraint for each individual traffic would be useful for streaming traffic, and it remains open to find the optimal scheduling under such constraints.

4. For the noisy channel estimation case, we have shown that a suboptimal backoff scheme would achieve a significant performance improvement over non-adaptive schemes. It remains open to investigate the optimal scheduling scheme when the noisy channel estimation can be refined by sending a second channel probing packet. With two-level incomplete information about the channel condition, we expect that the optimal scheduling would achieve better performance than that with only one-level information.

5. We note that in OMC-MAC, the channels may be under-utilized when the traffic load is low. In this case, it is plausible to enhance the bandwidth utilization by adapting the channel usage based on the traffic conditions. Observe that a node can estimate the traffic load by monitoring its FCL list and the channel contention outcomes. If the traffic load is low, this node can choose to utilize several idle channels simultaneously for data transmissions assuming that multiple channels can be used concurrently.

8.3.2. Utility-based Random Access and Flow Control

The impact of stochastic noisy feedback on distributed algorithms is an important yet under-explored area. Besides the issues studied in this dissertation, there are also many other challenges towards a complete theory under stochastic noisy feedback, for example, when we have non-convex problem formulations or when there exists asynchronism. An even more fundamental issue is when the network utility maximization approach would be robust and fast-converging in multi-hop wireless networks, given the noisy and lossy nature of wireless communications, and which decomposition and distributed algorithm among the alternatives has the best robustness property. All these questions remain open.

Another interesting question to ask is that how TCP flow control algorithms work in concert with opportunistic scheduling algorithms in wireless networks. Putting them in a complete picture requires a through investigation of joint flow control and opportunistic scheduling.

REFERENCES

[1] S. Adireddy and L. Tong, "Exploiting decentralized channel state information for random access," *IEEE Trans. Info. Theory*, vol. 51, no. 2, pp. 537–561, Feb. 2005.

[2] M. Adler, J.-Y. Cai, J. K. Shapiro, and D. Towsley, "Estimation of probabilistic congestion price using packet marking," in *Proceedings of IEEE INFOCOM'03*, San Francisco, CA, 2003.

[3] R. Agrawal, A. Bedekar, R. J. La, R. Pazhyannur, and V. Subramanian, "Class and channel condition based scheduler for EDGE/GPRS," in *Proceeding of SPIE*, 2001.

[4] D. Aguayo, J. Bicket, S. Biswas, G. Judd, and R. Morris, "Link-level measurements from an 802.11b mesh network," 2004.

[5] M. Andrews, "Joint optimization of scheduling and congestion control in communication networks," in *Proc. of CISS*, 2006.

[6] M. Andrews, K. Kumaran, K. Ramanan, A. Stolyar, P. Whiting, and R. Vijayakumar, "Providing quality of service over a shared wireless link," *IEEE Comm. Magazine*, vol. 39, pp. 150–154, 2001.

[7] M. Arulampalam, S. Maskell, N. Gordon, and T. Clapp, "A tutorial on particle filters for online nonlinear/non-gaussian bayesian tracking," *IEEE Trans. On Signal Processing*, vol. 50, no. 2, Feb. 2002.

[8] S. Athuraliya, V. H. Li, S. H. Low, and Q. Yin, "REM: Active queue management," *IEEE Network*, vol. 15, pp. 48–53, May/June 2001.

[9] F. Baccelli and D. Hong, "AIMD, fairness and fractal scaling of TCP traffic," in *Proc. of INFOCOM*, 2002.

[10] L. Bajaj, M. Takai, R. Ahuja, K. Tang, R. Bagrodia, and M. Gerla, "GloMoSim: A scalable network simulation environment," *Technical Report, UCLA Computer Science Department*.

[11] B. Bensaou, Y. Wang, and C. Ko, "Fair medium access in 802.11 based wireless ad-hoc networks," in *Proceedings of MOBICOM*, 2000.

[12] A. Benveniste, M. Metivier, and P. Priouret, *Adaptive Algorithms and Stochastic Approximation*. Springer, 1999.

[13] D. P. Bertsekas, *Nonlinear Programming*. Belmont, MA: Athena Scientific, 1995.

[14] D. P. Bertsekas and R. Gallager, *Data Networks*. New Jersey: Prentice Hall, 1992.

[15] D. P. Bertsekas and J. N. Tsitsiklis, *Parallel and Distributed Computation: Numerical Methods*. Prentice-Hall, 1989.

[16] ——, *Neuro-Dynamic Programming*. Massachusetts: Athena Scientific, 1996.

[17] V. Bharghavan, A. Demers, S. Shenker, and L. Zhang, "MACAW: A media acess protocol for wireless LAN's," in *Proceedings of SIGCOMM*, 1994.

[18] G. Bianchi, "Performance analysis of the IEEE 802.11 distributed coordination function," *IEEE JSAC*, vol. 18, March 2000.

[19] G. Bianchi, L. Fratta, and M. Oliveri, "Performance evaluation and enhancement of the csma/ca mac protocol for IEEE 802.11 wireless lans," in *PIMRC'96*, 1996.

[20] G. Bianchi and I. Tinnirello, "Kalman filter estimation of the number of competing terminals in an ieee 802.11 network," in *INFOCOM'03*, 2003.

[21] P. Billingsley, *Probability and Measure*, 3rd ed. John Wiley & Sons, Inc., 1995.

[22] T. Bonald and L. Massoulie, "Impact of fairness on internet performance," in *Proc. Sigmetrics*, 2001.

[23] V. Borkar, "Stochastic approximation with two time scales," *Systems and Control Letters*, vol. 29, pp. 291–294, 1997.

[24] S. Borst, "User-level performance of channel-aware scheduling algotirthms in wireless data networks," in *Proc. IEEE INFOCOM'03*, 2003.

[25] S. Borst and P. Whiting, "Dynamic rate control algorithms for HDR throughput optimization," in *Proc. IEEE INFOCOM'01*, pp. 976–985.

[26] R. Buche and H. J. Kushner, "Rate of convergence for constrained stochastic approximation algorithms," *SIAM Journal on Control and Optimization*, vol. 40, pp. 1011–1041, 2001.

[27] F. Cali, M. Conti, and E. Gregori, "Dynamic tuning of the IEEE 802.11 protocol to achieve a theoretical throughput limit," *IEEE/ACM Transactions on Networking*, vol. 8, no. 6, 2000.

[28] M. Cao, V. Raghunathan, and P. R. Kumar, "Cross layer exploitation of MAC layer diversity in wireless networks," in *Proceedings of IEEE ICNP*, 2006.

[29] C. S. Chang and Z. Liu, "A bandwidth sharing theory for a large number of HTTP-like connections," *IEEE/ACM Trans. Networking*, vol. 12, no. 5, pp. 952–962, Oct. 2004.

[30] N. Chang and M. Liu, "Optimal channel probing and transmission scheduling in a multichannel system," in *Proceedings of the Second Workshop on Information Theory and Applications*, 2007.

[31] P. Chaporkar and S. Sarkar, "Wireless multicast: Theory and approaches," *IEEE Trans. on Information Theory*, vol. 51, no. 6, pp. 1954–1972, June 2005.

[32] L. Chen, S. H. Low, M. Chiang, and J. C. Doyle, "Jointly optimal congestion control, routing, and scheduling for wireless ad hoc networks," in *Proc. IEEE INFOCOM'06*, 2006.

[33] M. Chiang, "Balancing transport and physical layers in wireless multihop networks: Jointly optimal congestion control and power control," *IEEE J. Sel. Areas Comm.*, vol. 23, no. 1, pp. 104–116, Jan. 2005.

[34] M. Chiang, S. H. Low, R. A. Calderbank, and J. C. Doyle, "Layering as optimization decomposition," *Proceedings of IEEE*, Jan. 2007.

[35] Y. S. Chow, H. Robbins, and D. Siegmund, *Great Expectations: Theory of Optimal Stopping.* Houghton Mifflin, 1971.

[36] S. Deb, S. Shakkottai, and R. Srikant, "Asymptotic behavior of internet congestion controllers in a many-flow regime," *Math. Operations Research*, 2005.

[37] A. Eryilmaz and R. Srikant, "Joint congestion control, routing and MAC for stability and fairness in wireless networks," in *Proceedings of IZS'06*, ETH Zurich, Switzerland, Feb. 2006.

[38] Z. Fang and B. Bensaou, "Fair bandwidth sharing algorithms based on game theory frameworks for wireless ad-hoc networks," in *Proceedings of INFOCOM*, 2004.

[39] T. Ferguson, *Optimal Stopping and Applications.* http://www.math.ucla.edu/~tom/Stopping/Contents.html, 2006.

[40] I. Gelfand and S. Fomin, *Calculus of Variations.* Prentice-Hall, 1963.

[41] S. Guha, K. Munagala, and S. Sarkar, "Jointly optimal transmission and probing strategies for multichannel wireless systems," in *Proceedings of CISS'06*, Princeton, NJ, 2006.

[42] P. Gupta and A. L. Stolyar, "Optimal throughput allocation in general random-access networks," in *Proc. CISS*, 2006.

[43] B. Hajek, "Stochastic approximation methods for decentralized control of multiaccess communications," *IEEE Transactions on Information Theory*, vol. 31, no. 2, pp. 176–184, Mar. 1985.

[44] G. Holland, N. Vaidya, and P. Bahl, "A rate-adaptive MAC protocol for multihop wireless networks," in *Proceedings of ACM/IEEE MOBICOM'01*, Rome, Italy, 2001.

[45] H. j. Kushner and J. Yang, "Analysis of adaptive step size sa algoithms for parameter tracking," *IEEE Trans. Automatic Control*, vol. 40, no. 8, pp. 1403–1410, 1995.

[46] N. Jain and S. Das, "A multichannel CSMA MAC protocol with receiver-based channel selection for multihop wireless networks," in *Proc. IC3N'01*, 2001.

[47] Z. Ji, Y. Yang, J. Zhou, M. Takai, and R. Bagrodia, "Exploiting medium access diversity in rate adaptive wireless LANs," in *Proceedings of MOBICOM'04*, 2004.

[48] T. Kailath, A. Sayed, and B. Hassibi, *Linear Estimation*. Prentice-Hall, 2000.

[49] A. Kamerman and L. Monteban, "WaveLAN-II: A high-performance wireless LAN for the unlicensed band," *Bell Labs Technical J.*, pp. 118–133, Summer 1997.

[50] K. Kar, S. Sarkar, and L. Tassiulas, "Achieving proportional fairness using local information in Aloha networks," *IEEE Transactions on Automatic Control*, 2004.

[51] F. P. Kelly, "Fairness and stability of end-to-end congestion control," *European Journal of Control*, pp. 159–176, 2003.

[52] F. P. Kelly, A. Maulloo, and D. K. H. Tan, "Rate control for communication networks: Shadow price, proportional fairness and stability," *Journal of Operational Research Society*, pp. 237–252, 1998.

[53] D. Kennedy, "On a constrained optimal stopping problem," *J. Appl. Prob.*, vol. 19, pp. 631–641, 1982.

[54] R. Knopp and P. Humlet, "Information capacity and power control in single cell multiuser communications," in *Proc. IEEE ICC 95*, June 1995.

[55] H. Kushner and G. Yin, *Stochastic Approximation and Recursive Algorithms and Applications*. Springer, 2003.

[56] J. W. Lee, M. Chiang, and R. A. Calderbank, "Utility-optimal random-access control," *IEEE Trans. Wireless Comm.*, vol. 6, no. 7, pp. 2741–2751, July 2007.

[57] X. Lin and N. Shroff, "The impact of imperfect scheduling on cross-layer rate control in multihop wireless networks," in *Proc. IEEE INFOCOM'05*, 2005.

[58] X. Lin and N. B. Shroff, "Utility maximization for communication networks with multipath routing," *IEEE Trans. Auto. Control*, vol. 51, no. 5, pp. 766–781, May 2006.

[59] X. Liu, E. K. Chong, and N. B. Shroff, "Opportunistic transmission scheduling with resource-sharing constraints in wireless networks," *IEEE JSAC*, vol. 19, no. 10, pp. 2053–2064, Oct. 2001.

[60] ——, "A framework for opportunistic scheduling in wireless networks," *Computer Networks*, vol. 41, no. 4, pp. 451–474, Mar. 2003.

[61] S. Low and R. Srikant, "A mathematical framework for designing a low-loss low-delay internet," *Network and Spatial Economics*, vol. 4, no. 1, pp. 75–101, 2004.

[62] S. Low, "Internet congestion control," *IEEE Control Systems Magazine*, pp. 28–43, 2002.

[63] S. Low and D. Lapsley, "Optimization flow control, I: Basic algorithm and convergence," *IEEE Trans. on Networking*, Dec. 1999.

[64] S. Low, L. Peterson, and L. Wang, "Understanding vegas: a duality model," *Journal of the ACM*, vol. 49, mar. 2002.

[65] S. Lu, T. Nandagopal, and V. Bharghavan, "A wireless fair service algorithm for packet cellular networks," in *Proceedings of MOBICOM*, 1998.

[66] H. J. Marquez, *Nonlinear Control Systems: Analysis and Design*. John Wiley & Sons, Inc., 2003.

[67] M. Mehyar, D. Spanos, and S. Low, "Optimization flow control with estimation error," in *Proc. INFOCOM'04*, 2004.

[68] J. Mo and J. Walrand, "Fair end-to-end window-based congestion control," *IEEE Trans. on Networking*, vol. 8, pp. 556–566, nov. 2000.

[69] T. Nandagopal, T.-E. Kim, X. Gao, and V. Bharghavan, "Achieving MAC layer fairness in wireless packet networks," in *MOBICOM*, 2000.

[70] A. Nasipuri, J. Zhuang, and S. Das, "A multichannel CSMA MAC protocol for multihop wireless networks," in *Proc. IEEE WCNC'99*, 1999.

[71] M. J. Neely, E. Modiano, and C. Li, "Fairness and optimal stochastic control for heterogeneous networks," in *Proc. of INFOCOM*, 2005.

[72] T. Ng, I. Stoica, and H. Zhang, "Packet fair queueing algorithms for wireless networks with location-dependent errors," in *IEEE INFOCOM'00*, 2000.

[73] T. Ozugur, M. Naghshineh, P. Kermani, C. Olsen, B. Rezvani, and J. Copeland, "Banlanced media access methods for wireless networks," in *MOBICOM'98*, 1998.

[74] D. Palomar and M. Chiang, "Alternative decomposition for network utility maximization: Framework and applications," in *INFOCOM'06*, 2006.

[75] ——, "A tutorial on decomposition methods and distributed network resource allocation," *IEEE J. Sel. Area Comm.*, vol. 24, no. 8, pp. 1439–1451, Aug. 2006.

[76] D. Qiao, S. Choi, and K. G. Shin, "Goodput analysis and link adaptation for IEEE 802.11a wireless LANs," *IEEE Trans. on Mobile Computing*, vol. 1, no. 4, pp. 278–292, 2002.

[77] X. Qin and R. Berry, "Exploiting multiuser dieversity for medium access control in wireless networks," in *Proceedings of IEEE INFOCOM'03*, 2003.

[78] S. Ray, J. Carruthers, and D. Starobinski, "RTS/CTS-induced congestion in ad-hoc wireless LANs," in *Proceedings of WCNC'03*, New Orleans, LA, 2003.

[79] S. I. Resnick, *A Probability Path*. Birkhauser, 1999.

[80] A. Sabharwal, A. Khoshnevis, and E. Knightly, "Opportunistic spectral usage: Bounds and a multi-band CSMA/CA protocol," *to appear IEEE/ACM Transactions on Networking*, 2006.

[81] B. Sadeghi, V. Kanodia, A. Sabharwal, and E. Knightly, "Opportunistic media access for multirate ad hoc networks," in *MOBICOM'02*, 2002.

[82] S. Shakkottai, R. Srikant, and A. L. Stolyar, "Pathwise optimality and state space collapse for the exponential rule," in *ISIT'02*, July 2002.

[83] A. N. Shiryayev, *Optimal Stopping Rules*. Springer-Verlag, 1978.

[84] J. So and N. Vaidya, "A multi-channel MAC protocol for ad hoc wireless networks," *Technical Report*, 2003.

[85] R. Srikant, *The Mathematics of Internet Congestion Control*. Birkhauser, 2004.

[86] A. L. Stolyar, "Maximizing queuing network utility subject to stability: Greedy primal-dual algorithm," *Queueing Systems*, vol. 50, no. 4, pp. 401–457, 2005.

[87] A. S. Tanenbaum, *Computer Networks*. Prentice Hall, 1996.

[88] A. Tang, J.-W. Lee, J. Huang, M. Chiang, and A. Calderbank, "Reverse engineering MAC," in *Proc. WiOpt'06*, 2006.

[89] Y. Tay and K. Chua, "A capacity analysis for the IEEE 802.11 mac protocol," *ACM/Baltzer Wireless Networks*, vol. 7, no. 2, March 2001.

[90] P. Tinnakornsrisuphap, R. La, and A. Makowski, "Modeling TCP traffic with session dynamics - many sources asymptotics under ECN/RED gateways," in *Proc. of the 18th International Teletraffic Congress*, 2003.

[91] N. Vaidya, P. Bahl, and S. Gupta, "Distributed fair scheduling in a wireless LAN," in *Proceedings of MOBICOM*, 2000.

[92] A. Vakili, M. Sharif, and B. Hassibi, "The effect of channel estimation error on the throughput of broadcast channels," in *Proceedings of IEEE ICASP*, 2006.

[93] P. Viswanath, D. N. Tse, and R. Laroia, "Opportunistic beamforming using dumb antennas," *IEEE Trans. Info. Theory*, vol. 48, no. 6, pp. 1277–1294, June 2002.

[94] X. Wang and K. Kar, "Cross-layer rate control for end-to-end proportional fairness in wireless networks with random access," in *Proc. MOBIHOC'05*, 2005.

[95] S.-L. Wu, C.-Y. Lin, Y.-C. Tseng, and J.-P. Sheu, "A new multichannel MAC protocol with on-demand channel assignment for multi-hop mobile ad hoc networks," in *Proc. IEEE WCNC'00*, 2000.

[96] Y. Xi and E. M. Yeh, "Optimal capacity allocation, routing, and congestion control in wireless networks," in *Proc. of ISIT*, 2006.

[97] J. Zhang, E. Chong, and I. Kontoyiannis, "Unified spatial diversity combining and power allocation for CDMA systems in multiple time-scale fading channels," *IEEE JSAC*, vol. 19, no. 7, pp. 1276–1288, July 2001.

[98] J. Zhang, D. Zheng, and M. Chiang, "The impact of stochastic noisy feedback on distributed network utility maximization," in *INFOCOM 2007*, 2007.

[99] ——, "The impact of stochastic noisy feedback on distributed network utility maximization in multi-hop wireless networks," *Submitted to IEEE Trans. on Information Theory*, 2007.

[100] J. Zhang and D. Zheng, "A stochastic primal-dual algorithm for joint flow control and MAC design in multi-hop wireless networks," in *Proc. of CISS'06*, 2006.

[101] D. Zheng, W. Ge, and J. Zhang, "Distributed opportunistic scheduling for ad-hoc communications: An optimal stopping approach," in *MobiHoc'07, to appear*, 2007.

[102] ——, "Distributed opportunistic scheduling for ad-hoc networks with random access: An optimal stopping approach," *Submitted to IEEE Trans. Information Theory*, 2007.

[103] D. Zheng and J. Zhang, "Protocol design and performance analysis of frequency-agile multi-channel medium access control," *IEEE Trans. Wireless Comm.*, vol. 5, no. 10, pp. 2887–2895, Oct. 2006.

[104] ——, "A two-phase utility maximization framework for wireless medium access control," *IEEE Trans. Wireless Comm., to appear*, Aug. 2007.

[105] D. Zheng, M. Cao, J. Zhang, and P. R. Kumar, "Channel aware distributed scheduling for exploiting multi-reciever diversity and multiuser diversity in ad-hoc networks: A unified PHY/MAC approach," in *Submitted to INFOCOM'08*.

APPENDIX A

SOME PROOFS FOR CHAPTER 4

A.1. Proof of Proposition 4.2.2

Proof: For $x > 0$, define the reward function as

$$Z_{n,j}(x) \triangleq R_{n,t(n),j}T - x\left[\sum_{i=1}^{n-1}[K_i\tau + (L-1)\tau] + K_n\tau + j\tau + T\right].$$

Suppose that a total of $\mathcal{K}$ receivers have been probed. It follows that

$$Z_{\mathcal{K}}(x) = Z_{\lceil\frac{\mathcal{K}}{L}\rceil,\text{mod}(\mathcal{K}-1,L)}(x).$$

To maximize the rate of return given by

$$x = \frac{E\left[R_{\lceil\frac{\mathcal{K}}{L}\rceil,tr(\lceil\frac{\mathcal{K}}{L}\rceil),\text{mod}(\mathcal{K}-1,L)} \times T\right]}{E\left[T_{\lceil\frac{\mathcal{K}}{L}\rceil,\text{mod}(\mathcal{K}-1,L)}\right]},$$

a key step is to find an optimal stopping algorithm $N^*(x)$ such that

$$V_0^*(x) = \sup_{N \in Q} E[Z_N(x)].$$

We note that the existence of $N^*(x)$ is guaranteed under the assumption **A1)** [102]. Once the optimal $N^*(x)$ and the corresponding $V_0^*(x)$ are found, the optimal throughput x_{SPWOR}^* can be obtained by solving $V_0^*(x) = 0$.

Observing that the problem is time invariant in n, we next characterize $N^*(x)$ using the backward induction method. To this end, suppose the transmitter has probed $L - 2$ receivers, and decides to continue probing receiver $L - 1$ at a further cost of $x\tau$. Assume that the instant rate for receiver $L - 1$ is $R_{n,t(n),L-1}$. Observe that if the transmitter decides to skip $(L-1)$th probed receiver, then all transmitters would re-contend for the channel. Accordingly, the highest expected reward that can be obtained for skipping the data transmission is

$$V_0^*(x) - x\left[\sum_{i=1}^{n}[K_i\tau + (L-1)\tau]\right].$$

The principle of optimality [39] dictates that the transmitter starts the data transmission to receiver $L - 1$ if

$$R_{n,t(n),L-1}T - xT_{n,L-1} \geq V_0^*(x) - x\left[\sum_{i=1}^n [K_i\tau + (L-1)\tau]\right],$$

which is equivalent to

$$R_{n,t(n),L-1} \geq \frac{V_0^*(x)}{T} + x. \tag{A.1}$$

Otherwise, it skips the data transmission, and the nodes re-contend for the channel.

It follows that the highest expected reward before the transmitter probes receiver $L - 1$, denoted as $V_{L-1}^*(x)$, is given by

$$V_{L-1}^*(x) = E[\max(RT - xT - x\tau, V_0^*(x) - x\tau)]. \tag{A.2}$$

Along the same line of reasoning, after receiver $L - 2$ is probed, the transmitter would transmit to receiver $L - 2$ if

$$R_{n,t(n),L-2}T - xT_{n,L-2} \geq V_{L-1}^*(x) - x(T_{n,L-2} - T)$$

which is equivalent to

$$R_{n,t(n),L-2} \geq \frac{V_{L-1}^*(x)}{T} + x. \tag{A.3}$$

Continue this procedure backward to receiver 0, and observe that $V_1^*(x)$ and $V_0^*(x)$ are related as follows:

$$V_0^*(x) = E[\max(RT - xT - xK\tau, V_1^*(x) - xK\tau)]. \tag{A.4}$$

Note that since $V_0^*(x_{SPWOR}^*) = 0$, the results does not change if we replace $V_j^*(x)$ with $V_j^*(x)/T$, for $j = 0, 1, \ldots, L - 1$. It follows that x_{SPWOR}^* is a solution to the fixed point equation (4.13), and (4.11) is an optimal stopping rule. To show that the fixed point equation has a unique solution x^*_{SPWOR}, we first note that $V^*_{L-1}(x) = E[(R-x)^+] - x\delta$ is continuous in x [102] and strictly decreases from $E[R]$ to $-\infty$ as x increases from 0 to ∞. Therefore, $V^*_{L-2}(x) = E[\max(R-x, v^*_{L-1}(x))] - x\delta$ is continuous and strictly decreases from $E[\max(R, E[R])]$ to $-\infty$ as x increases from 0 to ∞. It is clear that $E[\max(R, E[R])] \geq E[R] \geq 0$. Repeating the same procedure, it can be shown that $V^*_0(x)$ is a continuous function of x, and strictly decreases from a non-negative number to the negative infinity. We conclude that $V^*_0(x) = 0$ has a unique solution.

∎

A.2. Proof of Corollary 4.2.1

Proof: Clearly, based on (4.12), (4.16) is equivalent to

$$v^*_1 \geq v^*_2 \geq \cdots \geq v^*_{L-1} \geq v^*_L = 0. \tag{A.5}$$

Observe from (4.15) that

$$v^*_{L-1} = E[(R - x^*_{SPWOR})^+] - x^*_{SPWOR}\delta. \tag{A.6}$$

We first show that $v^*_{L-1} \geq 0$. Suppose that $v^*_{L-1} < 0$. Then, it follows that

$$v^*_{L-2} = E[\max(R - x^*_{SPWOR}, v^*_{L-1})] - x^*_{SPWOR}\delta$$
$$\leq E[(R - x^*_{SPWOR})^+] - x^*_{SPWOR}\delta = v^*_{L-1} < 0. \tag{A.7}$$

By induction, we can further show that

$$E[\max(R - x, V^*_1(x^*_{SPWOR}))] - \frac{x^*_{SPWOR}\delta}{p_s} < 0, \tag{A.8}$$

which contradicts (4.13). Therefore, $v^*_{L-1} \geq 0$.

It follows that

$$
\begin{aligned}
v^*_{L-2} &= E[\max(R - x^*_{SPWOR}, v^*_{L-1})] - x^*_{SPWOR}\delta \\
&\geq E[(R - x^*_{SPWOR})^+] - x^*_{SPWOR}\delta = v^*_{L-1}.
\end{aligned}
$$

Again, by induction, we conclude that $v^*_1 \geq v^*_2 \geq \cdots \geq v^*_{L-1} \geq v^*_L = 0$. ∎

A.3. Proof of Proposition 4.2.3

Proof: Define the reward function for SPWR probing as follows:

$$
Z_{n,j} \triangleq R_{n,j} - xT_{n,j}, \forall\, n = 0, 1, \ldots, \text{ and } j = 1, 2, \ldots, L,
$$

and $Z_K(x)$ is defined as

$$
Z_K(x) \triangleq Z_{\lceil \frac{K}{L} \rceil, \mathrm{mod}(K-1,L)}(x).
$$

The objective is to find the optimal stopping rule $N^*(x)$ such that $N^*(x) = \arg\sup_{N \in Q} E[Z_K(x)]$. It can be shown that $N^*(x)$ exists under conditions **A1**. Once we find the optimal $N^*(x)$ and

$$
U^*_0(x) = \sup_{N \in Q} E[Z_K(x)],
$$

the optimal throughput x^*_{SPWR} is then the solution to $U^*_0(x) = 0$.

Next, we derive the optimal stopping rule $N^*(x)$ using backward induction. Suppose that receiver $L - 1$ has been probed. If the transmitter $t(n)$ decides to skip the data transmission, then the best expected reward is given by $U^*_0(x) - x \left[\sum_{i=1}^n [K_i\tau + (L - 1)\tau]\right]$. Therefore, according to the principle of optimality, the transmitter will start the data transmission if

$$R_{n,L-1}T - xT_{n,L-1} \geq U_0^*(x) - x\left[\sum_{i=1}^{n}[K_i\tau + (L-1)\tau]\right],$$

or equivalently,

$$R_{n,L-1} \geq \frac{U_0^*(x)}{T} + x.$$

Otherwise, it skips the data transmission, and the transmitters would re-contend for the channel.

We next characterize the optimal stopping rule at receiver $j \in \{L-2, L-1, \ldots, 0\}$. To this end, define

$$U_{n,j}^*(x) \triangleq \operatorname*{ess\,sup}_{\mathcal{K} \geq (n-1)L+1+j} E[Z_{\mathcal{K}}|\mathcal{F}_{(n-1)L+1+j}], \forall\, n \geq 0, j \in \{0,1,\ldots,L-1\}, \quad (A.9)$$

where ess sup denotes the *essential supremum* [35, 39], and $\mathcal{F}_{(n-1)L+1+j}$ is the σ-algebra generated by the random variables up to the time $(n-1)L+1+j$.

To evaluate the stopping rule at receiver $L-2$, we first have that

$$U_{n,L-1}^*(x)$$

$$= \operatorname*{ess\,sup}_{\mathcal{K} \geq nL} E[Z_{\mathcal{K}}|\mathcal{F}_{nL}]$$

$$\overset{(a)}{=} \max\left\{R_{n,L-1}T - xT_{n,L-1}, \operatorname*{ess\,sup}_{\mathcal{K} \geq nL+1} E[Z_{\mathcal{K}}|\mathcal{F}_{nL}]\right\}$$

$$\overset{(b)}{=} \max\left\{R_{n,L-1}T - xT_{n,L-1}, \operatorname*{ess\,sup}_{\mathcal{K} \geq nL+1} E[Z_{\mathcal{K}-nL}|\mathcal{F}_{nL}] - x\sum_{i=1}^{n}[K_i\tau + (L-1)\tau]\right\}$$

$$\overset{(c)}{=} \max\left\{R_{n,L-1}T - xT_{n,L-1}, U_0^*(x) - x(T_{n,L-1} - T)\right\}$$

$$= \max\left\{R_{n,L-1}T, U_0^*(x) + xT\right\} - xT_{n,L-1},$$

where (a) is due to the property of the essential supremum [35, 39], (b) is from the property of conditional expectation, and (c) is due to that the problem is time invariant in n.

Therefore, using the principle of optimality, the transmitter transmits to receiver $L-2$ if

$$R_{n,L-2}T - xT_{n,L-2}$$

$$\geq E[U^*_{n,L-1}(x)|\mathcal{F}_{(n-1)L+1+L-2}]$$

$$= E[\max\{R_{n,L-1}T, U^*_0(x) + xT\} - xT_{n,L-1}|\mathcal{F}_{nL-1}]$$

$$= E[\max\{R_{n,L-2}T, R_{n,t(n),L-1}T, U^*_0(x) + xT\} - xT_{n,L-1}|\mathcal{F}_{nL-1}] \qquad \text{(A.10)}$$

After some algebra, (A.10) can be rewritten as

$$\psi^x_{L-2}(R_{n,L-2}) \leq 0, \qquad \text{(A.11)}$$

where $\psi^x_{L-2}(z)$ is defined as

$$\psi^x_{L-2}(z) \triangleq E\left[(\max\{R - z, U^*_0(x)/T + x - z\})^+\right] - x\delta. \qquad \text{(A.12)}$$

It is not difficult to see that $\psi^x_{L-2}(z)$ is a decreasing continuous function of z, and $\psi^x_{L-2}(z) \leq 0$ for sufficiently large z (note that x is positive). Therefore, from (A.11), the optimal stopping rule at receiver $L-2$ has the following threshold structure:

$$R_{n,L-2} \geq \theta_{L-2}(x), \qquad \text{(A.13)}$$

where

$$\theta_{L-2}(x) \triangleq \min\{z : \psi^x_{L-2}(z) \leq 0\}. \qquad \text{(A.14)}$$

Note that the minimum in (A.14) is achievable since $\{z : \psi^x_{L-2}(z) \leq 0\}$ is non-empty and $\psi^x_{L-2}(z)$ is continuous.

Repeating the same procedure from (A.10) to (A.14), one can show that the transmitter transmits to receiver $L-3$ if

$$\psi_{L-3}(R_{n,L-3}) \leq 0$$

where

$$\psi_{L-3}^x(z) \triangleq E\left[(\psi_{L-2}(\max(z, R)))^+ + (R - z)^+\right] - x\delta.$$

Since $\psi_{L-2}^x(z)$ is non-increasing and continuous, and $\psi_{L-2}^x(z) \leq 0$ for sufficiently large z, so is $\psi_{L-3}^x(z)$. Therefore, the optimal stopping rule at receiver $L - 3$ also has a threshold structure:

$$R_{n,L-3} \geq \theta_{L-3}(x),$$

where

$$\theta_{L-3}(x) \triangleq \min\{z : \psi_{L-3}^x(z) \leq 0\}. \qquad (A.15)$$

Similarly, it can be shown that the transmitter transmits to receiver j, for $j = L - 4, L - 5, \ldots, 0$, if

$$R_{n,j} \geq \theta_j(x),$$

where

$$\theta_j(x) = \min\{z : \psi_j(z) \leq 0\}, \qquad (A.16)$$

$$\psi_j^x(z) \triangleq E\left[(\psi_{j+1}(\max(z, R)))^+ + (R - z)^+\right] - x\delta. \qquad (A.17)$$

Again, based on the optimality equation, it can be shown that the optimal throughput x_{SPWR}^* is the solution to (4.23), and the uniqueness of the solution can be established along the same line as in the proof of Proposition 4.2.2. ∎

A.4. Proof of Corollary 4.2.2

Proof: The proof consists of three steps.

Step 1: We first show that $\theta^*_{L-3} \leq \theta^*_{L-2}$.

By the definition of θ^*_{L-2}, we have that

$$\psi^*_{L-2}(\theta^*_{L-2})$$

$$= E\left[\left(\max\left\{R - \theta^*_{L-2}, x^*_{SPWR} - \theta^*_{L-2}\right\}\right)^+\right] - x^*_{SPWR}\delta$$

$$= 0. \tag{A.18}$$

It follows from (4.22) that

$$\psi^*_{L-3}(\theta^*_{L-2})$$

$$= E\left[\left(\psi^*_{L-2}\left(\max(\theta^*_{L-2}, R)\right)\right)^+ + (R - \theta^*_{L-2})^+\right] - x^*_{SPWR}\delta$$

$$= E\left[(R - \theta^*_{L-2})^+\right] - x^*_{SPWR}\delta, \tag{A.19}$$

since $\psi^*_{L-2}\left(\max(\theta^*_{L-2}, R)\right) \leq 0$ by the definition of θ^*_{L-2}.

Comparing (A.18) with (A.19) yields that $\psi^*_{L-3}(\theta^*_{L-2}) \leq 0$. By the definition of θ^*_{L-3}, we have that $\theta^*_{L-3} \leq \theta^*_{L-2}$.

Step 2: To show that $\theta^*_0 \leq \theta^*_1 \leq \cdots \leq \theta^*_{L-3}$, we note that from (4.22), $\forall \, j = 0, 1, \ldots, L - 4$,

$$\psi^*_{j+1}(\theta^*_{j+1})$$

$$= E\left[\left(\psi^*_{j+2}\left(\max(\theta^*_{j+1}, R)\right)\right)^+ + (R - \theta^*_{j+1})^+\right] - x^*_{SPWR}\delta$$

$$= 0, \tag{A.20}$$

and

$$\psi^*_j(\theta^*_{j+1})$$

$$= E\left[\left(\psi^*_{j+1}\left(\max(\theta^*_{j+1}, R)\right)\right)^+ + (R - \theta^*_{j+1})^+\right] - x^*_{SPWR}\delta$$

$$= E\left[(R - \theta^*_{j+1})^+\right] - x^*_{SPWR}\delta. \tag{A.21}$$

Comparing (A.20) with (A.21) yields that $\psi_j^*(\theta_{j+1}^*) \leq 0$, and by the definition of θ_j^*, we have that $\theta_j^* \leq \theta_{j+1}^*$.

Step 3: We show that $\theta_{L-1}^* = x_{SPWR}^*$ has the smallest value among all the optimal thresholds.

We first prove that $\theta_{L-2}^* \geq \theta_{L-1}^*$ using contradiction. Suppose $\theta_{L-2}^* < \theta_{L-1}^*$. Combining with the results from **Step 1** and **2**, it immediately follows that $\theta_0^* \leq \theta_1^* \leq \cdots \leq \theta_{L-3}^* \leq \theta_{L-2}^* < \theta_{L-1}^*$, which simply indicates that in the optimal stopping rule for the SPWR mechanism, the transmitter never has to recall the previous probed receivers. However, this is the same mechanism as SPWOR probing, and by Corollary 4.2.1, the optimal thresholds $\{\theta_j^*, j = 0, 1, \ldots, L-1\}$ should be monotonically decreasing, leading to a contradiction to the above assumption. We conclude that $\theta_{L-2}^* \geq \theta_{L-1}^* = x_{SPWR}^*$.

It follows from (4.21) that

$$\psi_{L-2}^*(x_{SPWR}^*) = E\left[(R - x_{SPWR}^*)^+\right] - x_{SPWR}^*\delta \geq 0. \tag{A.22}$$

Next, we use (A.22) to show that $\theta_0^* \geq x_{SPWR}^*$. To see this, note that

$$\psi_0^*(x_{SPWR}^*)$$
$$= E\left[(\psi_1^*(\max(x_{SPWR}^*, R)))^+ + (R - x_{SPWR}^*)^+\right] - x_{SPWR}^*\delta$$
$$\geq E\left[(R - x_{SPWR}^*)^+\right] - x_{SPWR}^*\delta \geq 0. \tag{A.23}$$

By the definition of θ_0^*, we have that $\theta_0^* \geq \theta_{L-1}^* = x_{SPWR}^*$, and thereby concluding the proof. ∎

APPENDIX B

SOME PROOFS FOR CHAPTER 6

214

B.1. Proof of Theorem 7.3.1

Proof: Let $(\mathbf{x}^*, \boldsymbol{\lambda}^*)$ be a saddle point where $\mathbf{x}^*$ is the unique optimum solution to the primal problem and $\boldsymbol{\lambda}^* \in \Phi$. Define the Lyapunov function $V(\cdot, \cdot)$ as follows: [1]

$$V(\mathbf{x}, \boldsymbol{\lambda}) \triangleq ||\mathbf{x} - \mathbf{x}^*||^2 + \min_{\boldsymbol{\lambda}^* \in \Phi} ||\boldsymbol{\lambda} - \boldsymbol{\lambda}^*||^2, \tag{B.1}$$

and define the set

$$A_\mu \triangleq \{\{\mathbf{x}, \boldsymbol{\lambda}\} : V(\mathbf{x}, \boldsymbol{\lambda}) \leq \mu\} \tag{B.2}$$

for any given $\mu > 0$. In general, there can be multiple optimal shadow prices, meaning that

$$A_\mu = \bigcup_{\boldsymbol{\lambda}^* \in \Phi} \{\{\mathbf{x}, \boldsymbol{\lambda}\} : ||\mathbf{x} - \mathbf{x}^*||^2 + ||\boldsymbol{\lambda} - \boldsymbol{\lambda}^*||^2 \leq \mu\}.$$

A pictorial illustration of A_μ is provided in Fig. B.1.

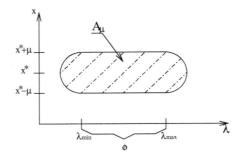

Figure B.1. A pictorial sketch of the set A_μ.

Step I: In what follows, we show that $\forall \mu > 0$, A_μ is recurrent, i.e., the iterates return to A_μ infinitely often with probability one.

Since $V(\cdot, \cdot)$ is continuous and Φ is compact, it follows that A_μ is compact and that the optimal shadow price is bounded above. Let $\lambda_0 > 0$ be an upper bound for the shadow

[1]Throughout $|| \cdot ||$ refers to the *Euclidean* norm.

price. For convenience, we define a "correction term" Z_n^λ to stand for the reflection effect that keeps the iterate $\lambda_l(n+1)$ in the constraint set. We rewrite (7.30) as

$$\lambda_l(n+1) = \lambda_l(n) - \epsilon_n[L_{\lambda_l}(\mathbf{x}(n), \boldsymbol{\lambda}(n)) + \beta_l(n) + \xi_l(n)] + \epsilon_n Z_n^{\lambda_l}, \qquad \text{(B.3)}$$

Similarly, define Z_n^x as the correction term for x_s, and accordingly,

$$x_s(n+1) = x_s(n) + \epsilon_n [L_{x_s}(\mathbf{x}(n), \boldsymbol{\lambda}(n)) + \alpha_s(n) + \zeta_s(n)] + \epsilon_n Z_n^{x_s}. \qquad \text{(B.4)}$$

Next, observe that

$$||\mathbf{x}(n+1) - \mathbf{x}^*||^2$$

$$\leq ||\mathbf{x}(n) - \mathbf{x}^*||^2 + \epsilon_n^2||L_\mathbf{x}(\mathbf{x}(n), \boldsymbol{\lambda}(n)) + \alpha(n) + \zeta(n)||^2$$

$$+ 2\epsilon_n(\mathbf{x}(n) - \mathbf{x}^*)^T [L_\mathbf{x}(\mathbf{x}(n), \boldsymbol{\lambda}(n)) + \alpha(n) + \zeta(n)]$$

where the inequality follows from the fact that the projection term $\epsilon_n Z_n^{x_s}$ is non-expansive [13]. Using the Cauchy-Schwartz Inequality, we have that

$$(\mathbf{x}(n) - \mathbf{x}^*)^T \alpha(n) \leq ||\mathbf{x}(n) - \mathbf{x}^*|| \cdot ||\alpha(n)||.$$

Recall that both $\{U_s\}$ and $\{f_l\}$ are twice-differentiable functions. It follows that the Lagrangian function $L(\mathbf{x}, \boldsymbol{\lambda})$ is also twice-differentiable and that both $L_x(\mathbf{x}(n), \boldsymbol{\lambda}(n))$ and $L_\lambda(\mathbf{x}(n), \boldsymbol{\lambda}(n))$ are bounded. Since $||\mathbf{x}(n) - \mathbf{x}^*||$ is bounded and $E_n(\zeta(n)) = 0$, combining Condition **A3** and the boundness of $\alpha(n)$ yields that

$$E_n(||\mathbf{x}(n+1)-\mathbf{x}^*||^2) \leq ||\mathbf{x}(n)-\mathbf{x}^*||^2+2\epsilon_n(\mathbf{x}(n)-\mathbf{x}^*)^T L_\mathbf{x}(\mathbf{x}(n), \boldsymbol{\lambda}(n))+O(\epsilon_n||\alpha(n)||)+O(\epsilon_n^2).$$

$$\text{(B.5)}$$

Along the same lines, we have that

$$E_n(||\boldsymbol{\lambda}(n+1)-\boldsymbol{\lambda}^*||^2) \leq ||\boldsymbol{\lambda}(n)-\boldsymbol{\lambda}^*||^2 - 2\epsilon_n(\boldsymbol{\lambda}(n)-\boldsymbol{\lambda}^*)^T L_\lambda(\mathbf{x}(n), \boldsymbol{\lambda}(n)) + O(\epsilon_n||\beta(n)||) + O(\epsilon_n^2).$$
(B.6)

Combining (B.5) and (B.6) yields that

$$E_n \left[||\mathbf{x}(n+1) - \mathbf{x}^*||^2 + ||\boldsymbol{\lambda}(n+1) - \boldsymbol{\lambda}^*||^2 \right] - \left(||\mathbf{x}(n) - \mathbf{x}^*||^2 + ||\boldsymbol{\lambda}(n) - \boldsymbol{\lambda}^*||^2 \right)$$

$$\leq 2\epsilon_n \underbrace{\left[(\mathbf{x}(n) - \mathbf{x}^*)^T L_\mathbf{x}(\mathbf{x}(n), \boldsymbol{\lambda}(n)) - (\boldsymbol{\lambda}(n) - \boldsymbol{\lambda}^*)^T L_\lambda(\mathbf{x}(n), \boldsymbol{\lambda}(n)) \right]}_{\triangleq G(\mathbf{x}(n), \boldsymbol{\lambda}(n))}$$
$$+ O\left(\epsilon_n(||\alpha(n)|| + ||\beta(n)||)\right) + O(\epsilon_n^2).$$
(B.7)

For convenience, define $\boldsymbol{\lambda}^*_{\min}(n) \triangleq \arg\min_{\boldsymbol{\lambda} \in \Phi} ||\boldsymbol{\lambda}(n) - \boldsymbol{\lambda}||^2$. Since Inequality (B.7) holds for every $\boldsymbol{\lambda}^* \in \Phi$, substituting $\boldsymbol{\lambda}^* = \boldsymbol{\lambda}^*_{\min}(n)$ yields that

$$E_n[V(\mathbf{x}(n+1), \boldsymbol{\lambda}(n+1))]$$

$$\leq E_n \left[||\mathbf{x}(n+1) - \mathbf{x}^*||^2 + ||\boldsymbol{\lambda}(n+1) - \boldsymbol{\lambda}^*_{\min}(n)||^2 \right]$$

$$\leq V(\mathbf{x}(n), \boldsymbol{\lambda}(n)) + 2\epsilon_n G(\mathbf{x}(n), \boldsymbol{\lambda}(n)) + O\left(\epsilon_n(||\alpha(n)|| + ||\beta(n)||)\right) + O(\epsilon_n^2), \text{(B.8)}$$

with the understanding that

$$G(\mathbf{x}(n), \boldsymbol{\lambda}(n)) = (\mathbf{x}(n) - \mathbf{x}^*)^T L_\mathbf{x}(\mathbf{x}(n), \boldsymbol{\lambda}(n)) - (\boldsymbol{\lambda}(n) - \boldsymbol{\lambda}^*_{\min}(n))^T L_\lambda(\mathbf{x}(n), \boldsymbol{\lambda}(n)).$$

Based on the structure of (B.8), we need the following lemma from [16] to establish the recurrence of A_μ.

Lemma 2.1.1. *(A Supermartingale Lemma) Let $\{X_n\}$ be an $\mathcal{R}^r$-valued stochastic process, and $V(\cdot)$ be a real-valued non-negative function in $\mathcal{R}^r$. Suppose that $\{Y_n\}$ is a* *sequence of random variables satisfying that* $\sum_n |Y_n| < \infty$ *with probability one. Let* $\{\mathcal{F}_n\}$ *be a sequence of* σ−*algebras generated by* $\{X_i, Y_i, i \leq n\}$. *Suppose that there exists a compact set* $A \subset \mathcal{R}^r$ *such that for all* n,

$$E_n[V(X_{n+1})] - V(X_n) \leq -\epsilon_n\delta + Y_n, \text{ for } X_n \notin A, \tag{B.9}$$

where ϵ_n *satisfies* **A1** *and* δ *is a positive constant. Then the set* A *is recurrent for* $\{X_n\}$, *i.e.,* $X_n \in A$ *for infinitely many* n *with probability one.*

Appealing to Lemma 2.1.1, it suffices to show that $G(\mathbf{x}(n), \boldsymbol{\lambda}(n)) < 0$ for $(\mathbf{x}(n), \boldsymbol{\lambda}(n)) \in A_\mu^c$. Since the Lagrangian function $L(\mathbf{x}(n), \boldsymbol{\lambda}(n))$ is concave in $\mathbf{x}$, it follows that

$$L(\mathbf{x}(n), \boldsymbol{\lambda}(n)) - L(\mathbf{x}^*, \boldsymbol{\lambda}(n)) \geq (\mathbf{x}(n) - \mathbf{x}^*)^T L_\mathbf{x}(\mathbf{x}(n), \boldsymbol{\lambda}(n)). \tag{B.10}$$

Since $L(\mathbf{x}, \boldsymbol{\lambda})$ is linear in $\boldsymbol{\lambda}$, we have that

$$L(\mathbf{x}(n), \boldsymbol{\lambda}_{\min}^*) - L(\mathbf{x}(n), \boldsymbol{\lambda}(n)) = (\boldsymbol{\lambda}_{\min}^* - \boldsymbol{\lambda}(n))^T L_\lambda(\mathbf{x}(n), \boldsymbol{\lambda}(n)). \tag{B.11}$$

Combining (B.10) and (B.11) yields that

$$
\begin{aligned}
G(\mathbf{x}(n), \boldsymbol{\lambda}(n)) \quad &\leq \quad L(\mathbf{x}(n), \boldsymbol{\lambda}_{\min}^*) - L(\mathbf{x}^*, \boldsymbol{\lambda}(n)) \\
&= \quad L(\mathbf{x}(n), \boldsymbol{\lambda}_{\min}^*) - L(\mathbf{x}^*, \boldsymbol{\lambda}_{\min}^*) \tag{B.12} \\
&\quad + L(\mathbf{x}^*, \boldsymbol{\lambda}_{\min}^*) - L(\mathbf{x}^*, \boldsymbol{\lambda}(n)). \tag{B.13}
\end{aligned}
$$

Since $(\mathbf{x}^*, \boldsymbol{\lambda}^*)$ is a saddle point, it follows that

$$L(\mathbf{x}(n), \boldsymbol{\lambda}^*) \leq L(\mathbf{x}^*, \boldsymbol{\lambda}^*) \leq L(\mathbf{x}^*, \boldsymbol{\lambda}(n)),$$

indicating that both (B.12) and (B.13) are non-positive. Furthermore, there exists $\delta_\mu > 0$ such that $G(\{\mathbf{x}(n), \boldsymbol{\lambda}(n)\}) < -\delta_\mu$ when $(\mathbf{x}(n), \boldsymbol{\lambda}(n)) \in A_\mu^c$, where A_μ^c is the complement set of A_μ.

218

Summarizing, we conclude that $\{\mathbf{x}(n), \boldsymbol{\lambda}(n)\}$ return to A_μ infinitely often with probability one.

Step II: Next, we use local analysis to show that $\{(\mathbf{x}(n), \boldsymbol{\lambda}(n)\}), n = 1, 2, \ldots$ leaves $A_{3\mu}$ only finitely often with probability one. Let $\{n_k, k = 1, 2 \ldots\}$ denote the recurrent times with $(\mathbf{x}(n_k), \boldsymbol{\lambda}(n_k)) \in A_\mu$. It suffices to show that there exists $n_{k_0} \in \{n_k, k = 1, 2 \ldots\}$, such that for all $n \geq n_{k_0}$, the original iterates $\{(\mathbf{x}(n), \boldsymbol{\lambda}(n)), n = 1, 2, \ldots\}$ reside in $A_{3\mu}$ almost surely. The basic idea of Step II is depicted in Fig. B.2.

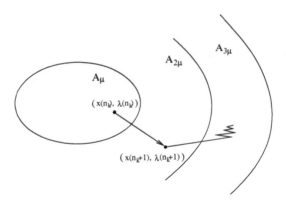

Figure B.2. A sketch of the basic idea in Step II.

Specifically, rewrite the algorithm as:

$$
\begin{bmatrix} x_s(n+1) \\ \lambda_l(n+1) \end{bmatrix} = \begin{bmatrix} x_s(n) \\ \lambda_l(n) \end{bmatrix} + \epsilon_n \begin{bmatrix} L_{x_s}(\mathbf{x}(n), \boldsymbol{\lambda}(n)) \\ -L_{\lambda_l}(\mathbf{x}(n), \boldsymbol{\lambda}(n)) \end{bmatrix} + \epsilon_n \begin{bmatrix} \alpha_s(n) + \zeta_s(n) \\ \beta_l(n) + \xi_l(n) \end{bmatrix} + \epsilon_n \begin{bmatrix} Z_n^{x_s} \\ Z_n^{\lambda_l} \end{bmatrix}.
$$
(B.14)

Observe that

1. Combining Condition **A1** and the boundedness of $L_x(\cdot, \cdot)$ and $L_\lambda(\cdot, \cdot)$, yields that
   $\epsilon_n L_x(\mathbf{x}(n_k), \boldsymbol{\lambda}(n_k)) \to 0$ and $\epsilon_n L_\lambda(\mathbf{x}(n_k), \boldsymbol{\lambda}(n_k)) \to 0$.

2. Condition **A2** implies that $\epsilon_n \alpha_s(n) \to 0$ and $\epsilon_n \beta_l(n) \to 0$.

3. Using Chebyshev's Inequality, for any small positive ρ, we have that

$$P(|\epsilon_n \zeta_s(n)| > \rho) \leq \epsilon_n^2 E[\zeta_s(n)^2]/\rho^2. \tag{B.15}$$

Based on Conditions **A1** and **A4**, we use the Borel-Cantelli Lemma to conclude that $\epsilon_n \zeta_s(n) > \rho$ only finitely often with probability one [21]. The same result can be obtained for $\epsilon_n \xi_l(n)$.

Combining the above observations with the properties of the reflection terms, we conclude that there exists n_{k_1} such that for all $n_k \geq n_{k_1}$, the overall change in one single step is no greater than μ and therefore the $(n_k + 1)$th iterate $(\mathbf{x}(n_k + 1), \boldsymbol{\lambda}(n_k + 1))$ resides in $A_{2\mu}$ with probability one.

Next, we show that there exists $n_{k_0} \geq n_{k_1}$ such that for all $n \geq n_{k_0}$, $\{\mathbf{x}(n), \boldsymbol{\lambda}(n)\} \in A_{3\mu}$ almost surely. Using (B.14), it suffices to show that the combined accumulative effect of

$$\epsilon_n \begin{bmatrix} L_{x_s}(\mathbf{x}(n), \boldsymbol{\lambda}(n)) \\ -L_{\lambda_l}(\mathbf{x}(n), \boldsymbol{\lambda}(n)) \end{bmatrix} \quad \text{and} \quad \epsilon_n \begin{bmatrix} \alpha_s(n) + \zeta_s(n) \\ \beta_l(n) + \xi_l(n) \end{bmatrix} \quad \text{cannot "drive" } \{\mathbf{x}(n), \boldsymbol{\lambda}(n)\} \text{ out of } A_{3\mu}$$

for sufficiently large n_{k_0}.

From Step I, it is clear that $\begin{bmatrix} L_{x_s}(\mathbf{x}(n), \boldsymbol{\lambda}(n)) \\ -L_{\lambda_l}(\mathbf{x}(n), \boldsymbol{\lambda}(n)) \end{bmatrix}$ is the gradient and would drive $(\mathbf{x}(n), \boldsymbol{\lambda}(n))$ towards A_μ because of the gradient descent properties. Furthermore, Condition **A2** reveals that $\lim_{m \to \infty} \sum_{n=m}^{\infty} \epsilon_n \alpha_s(n) \to 0$ and $\lim_{m \to \infty} \sum_{n=m}^{\infty} \epsilon_n \beta_l(n) \to 0$. Hence, the only possible terms that can drive $(\mathbf{x}(n), \boldsymbol{\lambda}(n))$ out of $A_{3\mu}$ are the martingale difference noises $\epsilon_n \zeta_s(n)$ and $\epsilon_n \xi_l(n)$. However, it follows from the Martingale Inequality [55] that

$$P\left\{ \sup_{k \geq m} | \sum_{n=m}^{k-1} \epsilon_n \zeta_s(n)| \geq \rho \right\} \leq \frac{\limsup_n E[\zeta_s(n)^2]}{\rho^2} \sum_{n=m}^{\infty} \epsilon_n^2,$$

and hence,

$$\lim_{m \to \infty} P \left\{ \sup_{k \ge m} | \sum_{n=m}^{k-1} \epsilon_n \zeta_s(n) | \ge \rho \right\} = 0, \forall \, \rho > 0. \tag{B.16}$$

That is to say, $\sup_{k \ge m} | \sum_{n=m}^{k-1} \epsilon_n \zeta_s(n) | \xrightarrow{P} 0$ as $m \to \infty$. It can be further shown that this is

equivalent to $\sup_{k \ge m} | \sum_{n=m}^{k-1} \epsilon_n \zeta_s(n) | \to 0$ almost surely (cf. Theorem 7.3.2 in [79]). Similarly,

$\limsup_{m \to \infty} | \sum_{n=m}^{\infty} \epsilon_n \xi_l(n) | = 0$ almost surely. In a nushell, these martingale difference noises

cannot drive $(\mathbf{x}(n), \boldsymbol{\lambda}(n))$ out of $A_{3\mu}$ for n_k sufficiently large.

Combining the above steps, we conclude that $\{(\mathbf{x}(n), \boldsymbol{\lambda}(n)), n = 1, 2, \ldots\}$ leaves $A_{3\mu}$

only finite often almost surely.

Since μ can be made arbitrarily small, it follows that $\{(\mathbf{x}(n), \boldsymbol{\lambda}(n),), n = 1, 2, \ldots\}$

converges w.p.1 to the optimal solutions. This concludes the proof.

∎

B.2. Proof of Theorem 7.3.2

Proof: Theorem 7.3.2 can be proven along the same line as that of Theorem 7.3.1 above.

In what follows, we outline a few key steps.

Define $A_{\mu,\eta} \triangleq A_\mu \cup A_\eta$. It can be shown that $A_{\mu,\eta}$ is compact. Following the same line

as in the proof for Theorem 7.3.1, it can be shown that

$$E_n[V(\mathbf{x}(n+1), \boldsymbol{\lambda}(n+1))] \le V(\mathbf{x}(n), \boldsymbol{\lambda}(n)) + 2\epsilon_n(1 - \eta)G(\mathbf{x}(n), \boldsymbol{\lambda}(n)) + O(\epsilon_n^2), \tag{B.17}$$

where $V(\cdot, \cdot)$ and $G(\cdot, \cdot)$ are defined in Appendix B.1. Recall that $G(\mathbf{x}(n), \boldsymbol{\lambda}(n)) < -\delta$ for

some positive constant δ when $(\mathbf{x}(n), \boldsymbol{\lambda}(n)) \in A_\mu^c$. Since $A_{\mu,\eta}^c$ is a subset of A_μ^c, it follows

that $G(\mathbf{x}(n), \boldsymbol{\lambda}(n)) < -\delta$ for $(\mathbf{x}(n), \boldsymbol{\lambda}(n)) \in A_{\mu,\eta}^c$. Accordingly, appealing to Lemma 2.1.1,

we conclude that the iterates return to $A_{\mu,\eta}$ infinitely often with probability one. Letting

$\mu \to 0$, we have that $A_{\mu,\eta} \to A_\eta$, thereby concluding the proof. ∎

B.3. Proof of Theorem 7.5.3

Proof: To establish the stability of the stochastic two time-scale algorithm based on primal decomposition, it suffices to show that on the faster time scale, the iterates $\boldsymbol{\lambda}(n)$ follow asymptotically the trajectory of the mean ODE in (7.40) and (7.41) with fixed **p**, and that on the larger time scale, the iterates $\mathbf{p}(n)$ follow the trajectory of the mean ODE (7.42) in the sub-gradient form. We note that due to the structure of this two time-scale algorithm, the establishment of the "passage" from the stochastic form to the mean ODEs is nontrivial, and it is intimately tied to the "amplified" noise in the gradient estimation and the curvatures of the utility functions.

As is standard [55], we define the continuous-time interpolation of $\{\boldsymbol{\lambda}(n)\}$ as

$$
\boldsymbol{\lambda}^0(t) = \begin{cases} \boldsymbol{\lambda}(0), & \text{if } t \le 0 \\[2mm] \boldsymbol{\lambda}(n), & \text{if } t_n^a \le t < t_{n+1}^a, \end{cases} \tag{B.18}
$$

and define the shifted process $\boldsymbol{\lambda}^n(t) \triangleq \boldsymbol{\lambda}^0(t_n + t)$ (see Fig. B.3). Define the interpolated (and shifted) process of **p** on the slower time scale analogously with $\mathbf{p}(n)$ in lieu of $\boldsymbol{\lambda}$.

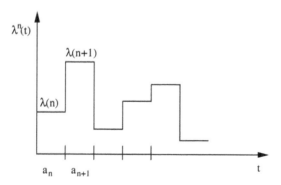

Figure B.3. An illustration of the shifted interpolated process of $\{\boldsymbol{\lambda}(n)\}$

We first show that on the faster time scale, the change of the shifted process $\mathbf{p}^n(t)$ is infinitesimal for any given t as $n \to \infty$. To see this, note that

$$\mathbf{p}^n(t) = \mathbf{p}(n) + \sum_{i=n}^{m^a(t+t_n)-1} b_i \mathbf{F}(i) + \sum_{i=n}^{m^a(t+t_n)-1} b_i \mathbf{N}(i),$$

where $\mathbf{F}_l(n) = \sum_{k \in L} \lambda_k^* \frac{\partial h_k(\mathbf{p}(n))}{\partial p_l}$. Thus, it suffices to show that the second term and the third term on the right hand side go to zero with probability one as $n \to \infty$ [55]. To this end, it is easy to see that the second term goes to zero since $\mathbf{F}$ is bounded above and that

$$\sum_{i=n}^{m^a(t+t_n)-1} b_i \leq \sup_{i \geq n} \frac{b_i}{a_i} \sum_{i=n}^{m^a(t+t_n)-1} a_i \stackrel{(A)}{\leq} t \sup_{i \geq n} \frac{b_i}{a_i} \stackrel{(B)}{\underset{n \to \infty}{\longrightarrow}} 0, \tag{B.19}$$

where (A) is based on the definition of $m^a(t)$ and (B) follows from Condition **B1**. To examine the third term, observe that

$$\sum_{i=n}^{m^a(t+t_n)-1} b_i \mathbf{N}(i) = \mathbf{N}^b(T_{n,1}) - \mathbf{N}^b(T_{n,2}), \tag{B.20}$$

where $m^b(T_{n,1}) = m^a(t+t_n)$ and $m^b(T_{n,2}) = m^a(t_n)$. Based on the definition of $m^b(t)$, we conclude that as $n \to \infty$, $T_{n,i} \to \infty$, $i = 1, 2$, and that

$$T_{n,1} - T_{n,2} < \sum_{i=m^a(t_n)}^{m^a(t+t_n)-1} b_i < \sum_{i=m^a(t_n)}^{m^a(t+t_n)-1} a_i \leq t.$$

It then follows from Condition **B2** that the third term goes to zero for any given t (note that in the above analysis, we have used the fact that if **B2** is valid for some positive T, then **B2** holds for all positive T [55].).

In summary, we have shown that on the faster time scale $\mathbf{p}^n(t)$ remains almost unchanged for any given t asymptotically. Next, we show that the shifted process $\boldsymbol{\lambda}^n(t)$ is equicontinuous in the extended sense and hence by the Arzela-Ascoli Theorem it contains a convergent sub-sequence whose limit follows the trajectory of the dual algorithm in

(7.40) and (7.41). Specifically, rewrite $\lambda_l^n(t)$ as follows:

$$
\begin{aligned}
&\lambda_l^n(t) \\
&= \lambda_l(n) - \sum_{i=n}^{m^a(t+t_n)} a_i \left[c_l(\mathbf{p}(i)) - \sum_{s \in \mathcal{S}(l)} U_s'^{-1} \left(\sum_{l \in \mathcal{L}(s)} \lambda_l(i) \right) \right] \\
&\quad + \sum_{i=n}^{m^a(t+t_n)} a_i \sum_{s \in \mathcal{S}(l)} U_s'^{-1} \left(\sum_{l \in \mathcal{L}(s)} \lambda_l(i) + M_s^x(i) \right) \\
&\quad - \sum_{i=n}^{m^a(t+t_n)} a_i \sum_{s \in \mathcal{S}(l)} U_s'^{-1} \left(\sum_{l \in \mathcal{L}(s)} \lambda_l(i) \right) \\
&\quad - \sum_{i=n}^{m^a(t+t_n)} a_i M_l^\lambda(i) + \sum_{i=n}^{m^a(t+t_n)} a_i Z_l^\lambda(i),
\end{aligned}
\tag{B.21}
$$

where $Z_l^\lambda(n)$ is the reflection term that keeps the shadow price non-negative. Observe that

$$
\begin{aligned}
&\left| \sum_{i=n}^{m^a(t+t_n)} a_i \sum_{s \in \mathcal{S}(l)} U_s'^{-1} \left(\sum_{l \in \mathcal{L}(s)} \lambda_l(i) + M_s^x(i) \right) - \sum_{i=n}^{m^a(t+t_n)} a_i \sum_{s \in \mathcal{S}(l)} U_s'^{-1} \left(\sum_{l \in \mathcal{L}(s)} \lambda_l(i) \right) \right| \\
&\stackrel{(A)}{=} \left| \sum_{s \in \mathcal{S}(l)} \sum_{i=n}^{m^a(t+t_n)} a_i \vartheta_{s,i} M_s^x(i) \right| \stackrel{(B)}{\longrightarrow} 0 \; w.p.1,
\end{aligned}
\tag{B.22}
$$

where (A) follows from the Mean Value Theorem and Condition **B3** (which indicates that $0 < \bar{\mu}_s \leq \vartheta_{s,i} \leq \bar{\nu}_s < \infty, \; \forall \, i, s$), and (B) from Condition **B2**. Again, based on Condition **B2**, we have that $\left| \sum_{i=n}^{m^a(t+t_n)} a_i M_l^\lambda(i) \right| \longrightarrow 0$. It then can be shown that $\sum_{i=n}^{m^a(t+t_n)} a_i \left[c_l(\mathbf{p}(i)) - \sum_{s \in \mathcal{S}(l)} U_s'^{-1} \left(\sum_{l \in \mathcal{L}(s)} \lambda_l(i) \right) \right]$ and $\sum_{i=n}^{m^a(t+t_n)} a_i Z_l^\lambda(i)$ are equicontinuous in the extended sense [55]. As a result, the shifted process $\boldsymbol{\lambda}^n(t)$ is equicontinuous in the extended sense, and the limit of the convergent subsequence follows the trajectory of the mean ODE in (7.40) and (7.41).

Similarly, on the slower time scale, it can be shown that the interpolated process of $\mathbf{p}(n)$ follows the trajectory of the mean ODE in (7.42). This, together with the convergence property of the mean ODEs (7.40), (7.41) and (7.42), concludes the proof. $\blacksquare$